Geographies of British Modernity

RGS-IBG Book Series

The *Royal Geographical Society (with the Institute of British Geographers) Book Series* provides a forum for scholarly monographs and edited collections of academic papers at the leading edge of research in human and physical geography. The volumes are intended to make significant contributions to the field in which they lie, and to be written in a manner accessible to the wider community of academic geographers. Some volumes will disseminate current geographical research reported at conferences or sessions convened by Research Groups of the Society. Some will be edited or authored by scholars from beyond the UK. All are designed to have an international readership and to both reflect and stimulate the best current research within geography.

The books will stand out in terms of:

* the quality of research
* their contribution to their research field
* their likelihood to stimulate other research
* being scholarly but accessible.

For series guides go to http://www.blackwellpublishing.com/pdf/rgsibg.pdf

Published

Forthcoming

Geographies of British Modernity

Space and Society in the Twentieth Century

Edited by
David Gilbert, David Matless
and Brian Short

Blackwell
Publishing

© 2003 by Blackwell Publishing Ltd

350 Main Street, Malden. MA 02148-5020, USA
108 Cowley Road, Oxford OX4 1JF, UK
550 Swanston Street, Carlton, Victoria 3053, Australia

The right of David Gilbert, David Matless and Brian Short to be identified as the Authors of
the Editorial Material in this Work has been asserted in accordance with the UK Copyright,
Designs, and Patents Act 1988.

First published 2003 by Blackwell Publishing Ltd

Library of Congress Cataloging-in-Publication Data

Geographies of British modernity : space and society in the twentieth century / edited by David
Gilbert, David Matless, and and Brian Short.
    p. cm. – (RGS-IBG book series)
Includes bibliographical references and index.
    ISBN 0–631–23500–0 (alk. paper)—ISBN 0–631–23501–9 (pbk.: alk. paper)
    1. Great Britain—Historical geography. 2. Great Britain—Social conditions—twentieth
century. 3. Great Britain—History—twentieth century. 4. Human geography–Great Britain.
I. Gilbert, David. II. Matless, David. III. Short, Brian, 1944-IV. Series.

DA600 .G45 2003
911'.41–dc21   2003000665

A catalogue record for this title is available from the British Library.

Set in 10/12 pt Plantin
by Kolam Information Services Pvt. Ltd, Pondicherry, India
Printed and bound in the United Kingdom
by MPG Books, Bodmin, Cornwall

For further information on
Blackwell Publishing, visit our website:
http://www.blackwellpublishing.com

Contents

Series Editors' Preface

The RGS–IBG Book series publishes the highest quality of research and scholarship across the broad disciplinary spectrum of geography. Addressing the vibrant agenda of theoretical debates and issues that characterize the contemporary discipline, contributions will provide a synthesis of research, teaching, theory and practice that both reflects and stimulates cutting-edge research. The series seeks to engage an international readership through the provision of scholarly, vivid and accessible texts.

Nick Henry and Jon Sadler
RGS–IGB Book Series Editors

Acknowledgements

The editors would like to thank the contributors to this volume for their cooperation, patience and prompt attention to our queries. A collaboration of this kind might have included many other authors and themes, and so we also thank those colleagues whose original proposals we finally had to omit because of shortage of space. Many colleagues have commented on whole or part of the proposal, and our thanks go to Alan Baker and Richard Dennis for their initial thoughts; to Nick Henry, Human Geography editor of the Blackwell IBG/RGS series, and to the anonymous referees whose reports were so thorough and helpful. We are also indebted to Jane Hotchkiss, Angela Cohen and Sarah Falkus from the Blackwell editorial team. The volume originated from the opportunity to meet the contributors at a Historical Geography Research Group symposium at the IBG/RGS 2000 conference at the University of Sussex, and we are grateful to the HGRG for its willingness to support our idea, and also to provide support for study group guests. Editorial meetings have taken our time away from departmental matters: we therefore thank our own colleagues at Nottingham, Royal Holloway and Sussex for their understanding; and the ideally located Pizza Paradiso in Store Street, London, for its ambience and stimulus to editorial thought.

We thank copyright holders for their kind permission to reproduce the images used. Every effort has been made to obtain permission to reproduce copyright material but if proper acknowledgement has not been made we would invite copyright holders to inform us of the oversight.

David Gilbert
David Matless
Brian Short

Contributors

Robin Butlin is Professor of Historical Geography at the University of Leeds. His research interests include historical geographies of: European imperialism, c.1880–1960, Palestine c.1600–1914, and English rural landscapes, together with the history of Historical Geography. His publications include *Historical Geography: Through the Gates of Space and Time* (1993); *Geography and Imperialism 1820–1940*, edited with Morag Bell and Michael Heffernan (1995); *Ecological Relations in Historical Times*, edited with N. Roberts (1995); *An Historical Geography of Europe*, edited with R. A. Dodgshon (1998); and *Place, Culture and Identity: Essays in Honour of Alan R. H. Baker*, edited with I. S. Black (2001).

Danny Dorling is Professor of Geography at the University of Leeds; he was previously Reader in Geography at the University of Bristol. His recent publications include *Health, Place and Society*, with M. Shaw and R. Mitchell (2000).

Bronwen Edwards is completing a Ph.D. at the London College of Fashion, London Institute. Her research concerns the architecture and culture of shopping in London's West End from 1930 to 1959. She has worked for the Twentieth Century Society and for English Heritage.

David Gilbert is a Senior Lecturer in the Department of Geography at Royal Holloway, University of London. His recent research has concerned the modern history of London, with particular emphasis on the development of suburban culture, and the influence of imperialism on London's physical and social landscapes. He was co-editor, with Felix Driver, of *Imperial Cities: Landscape, Display and Identity* (1999).

Gerry Kearns is a Lecturer in Geography at the University of Cambridge. His main research interests are the history of urban public health, the relations between geographic thought and imperialism, and the geographical dimensions of Irish identities. Recent publications relevant to the concerns of the chapter in this book include 'Time and Some Citizenship: Nationalism and Thomas Davis Bullán', *Irish Studies Journal*, 5 (2001), 23–54; and ' "Educate that holy hatred": Place, Trauma and Identity in the Irish Nationalism of John Mitchel', *Political Geography*, 20 (2001), 885–911.

Alun Howkins is Professor of Social History at the University of Sussex. He has published widely on the social and cultural history of rural England, and is the author of *Poor Labouring Men: Rural Radicalism in Norfolk, 1872–1923* (1985), and *Reshaping Rural England: A Social History, 1850–1925* (1992). He has recently published a substantial section of the Cambridge Agrarian History of England and Wales, volume 7, and as a regular broadcaster wrote and presented the major BBC series, *Fruitful Earth* in 1999. He is an editor of *History Workshop Journal* and *Oral History*.

Ron Johnston is a Professor in the School of Geographical Sciences at the University of Bristol; **Charles Pattie** is a Professor in the Department of Geography at the University of Sheffield; **Danny Dorling** is a Professor and **David Rossiter** is a Research Fellow in the School of Geography at the University of Leeds. They have collaborated widely in studies of the UK's electoral geography over the last decade, and have co-authored two recent books for Manchester University Press: *The Boundary Commissions: Redrawing the UK's Map of Parliamentary Constituencies* (1999), and *From Votes to Seats: The Operation of the UK Electoral System since 1945* (2001).

Dennis Linehan is a Lecturer in Cultural and Historical Geography at University College, Cork. He is currently working on a book on the question of industrial modernity and regionalism in Britain.

David Matless is Reader in Cultural Geography at the University of Nottingham. He is the author of *Landscape and Englishness* (1998) and editor of *The Place of Music* (1998). He is currently researching cultures of nature in twentieth-century England, and the life and work of ecologist and artist Marietta Pallis.

Peter Merriman is a Lecturer in the Department of Geography at the University of Reading. His research focuses on mobility, space and social theory, historical geographies of transport, and cultures of landscape in twentieth-century Britain. He is currently writing a book on the cultural geographies of the M1 motorway.

Simon Naylor is Lecturer in Human Geography in the School of Geographical Sciences, University of Bristol. He is a historical and cultural geographer with research interests in the geographies of science, technology and religion. He is co-editor, with Ian Cook, David Crouch and James Ryan, of *Cultural Turns/Geographical Turns: Perspectives on Cultural Geography* (2000).

Colin Pooley is Professor of Social and Historical Geography at Lancaster University. His research focuses on societal change in Britain and Europe over the last 200 years, and especially aspects of migration, mobility, health, housing and ethnicity. Publications include a wide range of journal articles, and his recent research with J. Turnbull on migration is reported in *Migration and Mobility in Britain since the Eighteenth Century* (1998).

Rebecca Preston is a researcher and writer on the history of gardening, and is currently Research Assistant at the AHRB Centre for the Study of the Domestic Interior at the Royal College of Art. Her Ph.D. at Royal Holloway, University of London on the culture and politics of gardening in nineteenth-century Britain was completed in 1999. Her published work explores the relationship between the practices of gardening and social identity.

James R. Ryan is Lecturer in Human Geography at Queen's University Belfast. He is author of *Picturing Empire* (1997) and co-editor, with Ian Cook, David Crouch and Simon Naylor, of *Cultural Turns/Geographical Turns: Perspectives on Cultural Geography* (2000), and, with Joan M. Schwartz, of *Picturing Place* (2003).

Brian Short is Professor of Historical Geography, and currently Head of the Department of Geography, at the University of Sussex. His research and publications have primarily focused on rural communities, economy and landscape, with a recent emphasis on the earlier twentieth century, as seen in his *Land and Society in Edwardian Britain* (1997) and *The National Farm Survey 1941–1943: State Surveillance and the Countryside in England and Wales in the Second World War,* co-authored with Charles Watkins, William Foot and Phil Kinsman (2000).

Chapter One

Historical Geographies of British Modernity

Brian Short, David Gilbert and
David Matless

Geographies of Modernity

In 1902 Heinemann of London published Halford Mackinder's *Britain and the British Seas*, a work of geographical synthesis in which 'the phenomena of topographical distribution relating to many classes of fact have been treated, but from a single standpoint and on a uniform method' (1902: vii). Reread a century later, Mackinder's account of Britain contains some elements that seem archaic and others that still appear perceptive or even visionary. One diagram in the book shows 'the relative nigrescence of the British population' (1902: 182). This used an index based on samples of hair colour to map the patterning of what were described as long-skulled and dark Celts, long-skulled and blond Teutons, and remnant groups of 'round-headed men'. Mackinder's diagram is now used only as a pedagogic device to illustrate the contemporary obsession with racial difference (and the shaky evidence on which it was based). Students familiar with the cultural complexities of early twenty-first-century Britain find the language, aims and methods of Mackinder's treatment of 'racial geography' perplexing, amusing, risible or offensive. In contrast, Mackinder's comments on the potential of tidal power as a replacement for fossil fuels, seem like a prophecy still waiting to be fulfilled:

> A vaster supply of energy than can be had from the coal of the whole world is to be found in the rise and fall of the tide upon the submerged plateau which is the foundation of Britain. No one has yet devised a satisfactory method of harnessing the tides, but the electrical conveyance of power has removed one at least of the impediments, and sooner or later, when the necessity is upon us, a way may be found of converting their rhythmic pulsation into electrical energy. (1902: 339)

It is appropriate to open this book on the *Geographies of British Modernity,* the first volume dedicated specifically to the historical geography of twentieth-century Britain, with reference to Mackinder, not just because *Britain and the British Seas* provides a convenient starting point from the early years of the century. As we argue later in this introductory essay, it is important to think about how the discipline of geography has changed and developed over the twentieth century as a way of writing about Britain and Britishness, and Mackinder is often credited with the establishment of British academic geography. But it is also appropriate to start with Mackinder because, as Gearóid Ó Tuathail (1992) has argued, his work as both academic geographer and as politician must be interpreted as a comment on the nature of the modern world and Britain's place within it. Mackinder is now best remembered as a geopolitician, through his theory of the 'world heartland' that influenced and legitimized American Cold War strategy. This work on the 'closed heart-land of Euro-Asia' as the 'geographical pivot of history' needs to be set within what was a much broader response to dramatic transformations that were taking place at the beginning of the twentieth century (Mackinder 1904: 434). Mackinder's work can be seen as an attempt 'to modernize traditional conservative myths about an organic community in an age where a multiplicity of international and domestic material transformations were eroding the economic foundations of the British Empire and the social world of the aristocratic establishment who ran it' (Ó Tuathail 1992: 102). Seen from this perspective, Mackinder's broader undertaking becomes a particular interpretation and projection of the geography of British modernity at the beginning of the twentieth century.

The term modernity is one for which there is certainly no simple or agreed definition: 'Its periodisations, geographies, characteristics and promise all remain elusive' (Ogborn 1998: 2). In general terms modernity has been seen as a description both of major social and material changes – particularly the emergence of the modern state, industrial capitalism, new forms of science and technology, and time-space compression – and of the growing consciousness of the novelty of these changes. This consciousness has been marked by pronounced double-sidedness or ambivalence. To be modern, in Marshall Berman's words, 'is to find ourselves in an environment that promises us adventure, power, joy, growth, transformation of ourselves and the world – and at the same time threatens to destroy everything we have, everything we know, everything we are' (1983: 15). Berman's account in *All That Is Solid Melts Into Air* is perhaps the most influential late twentieth-century interpretation of modernity. In a key passage he outlines the different dimensions of the 'creative destruction' of modernity:

> The maelstrom of modern life has been fed from many sources: great discoveries in the physical sciences, changing our images of the universe and our

place within it; the industrialisation of production which transforms scientific knowledge into technology, creates new human environments and destroys old ones, speeds up the whole tempo of life, generates new forms of corporate power and class struggle; immense demographic upheavals, severing millions of people from their ancestral habitats, hurtling them halfway across the world into new lives; rapid and often cataclysmic urban growth; systems of mass-communication, dynamic in their development, enveloping and binding together the most diverse people and societies; increasingly powerful nation states, bureaucratically structured and operated, constantly striving to expand their powers; mass social movements of people, and peoples, challenging their political and economic rulers, striving to gain some control over their lives; finally, bearing and driving all these people and institutions along an ever-expanding drastically fluctuating capitalist world market. (1983: 16)

As Berman acknowledges, this describes a vast history which is highly differentiated in time and space. The conservative imperialism of Mackinder and many of his contemporaries in the British establishment can be seen as a reaction both to the general characteristics of these changes at the beginning of the twentieth century, and to their specific impact upon Britain and the British empire. The beginning of the twentieth century was, as Stephen Kern (1983) has suggested, a time of sweeping change in technology and culture altering understandings of time, space and the nature of the world order. Significant technological innovations of the period included the telephone, wireless telegraphy, cinema, bicycle, automobile, and airplane, while contemporary cultural and intellectual developments included the emergence of psychoanalysis, cubism and relativity. The early twentieth century witnessed an acceleration in the rate of the 'internationalization of human affairs', a consequence in part of the time-space compression facilitated by new technology (Ó Tuathail 1992: 103). It also saw growing pressure in Western societies from groups previously marginalized – particularly the working classes and women – for greater political power and social justice.

We return to Mackinder and questions of British modernity below, but recent work has begun to ask specifically geographical questions about the nature of modernity *per se*. Historical geographers such as Ogborn (1998), Pred and Watts (1992) and Gregory (1994) have begun to demonstrate the impossibility of understanding modernity (or indeed any other historical formation) in an aspatial fashion, whether the concern is for the geographical project of empire, the spatial organization of industrial production, the relations of city and country, or the symbolic geographies of modern or anti-modern nationhood (see also Graham and Nash 2000). What we wish to emphasize in this collection is the spatial fabric of the modern in all the above senses, rather than geography being simply a fixed spatial background over which historical processes play. The understanding of modern times cannot achieve sufficiency apart from the understanding of modern spaces. In *Spaces of Modernity,* his account of 'London's Geographies 1680–1780', Miles Ogborn highlights three ways in which a geographical understanding may transform our sense of modernity: 'through investigating the forms in which the spatial is written into theories of modernity; by acknowledging the ways in which there are different modernities in different places; and by conceptualising modernity as a matter of the hybrid relationships and connections between places' (Ogborn 1998: 17). While this book is concerned with a very different period within 'modernity', these questions remain central, whether one is considering the rationalization of modern spaces through industrial policies or planning theories, the specifically British inflections of wider global processes, or ways in which local processes in twentieth-century Britain cannot be understood apart from imperial or postcolonial relationships.

If the geographies of modernity have been subject to various readings on different scales at different times, the geographies of *British* modernity have been less subjected to systematic analysis. Historians of twentieth-century Britain, and of the modernities of earlier British life, have developed sophisticated and often contrasting analyses of the ways in which Britain played a key role in the emergence of modernity *per se*, the particular inflections of the modern found in Britain as distinct from other Western powers, and the ways in which Britishness was imagined in relation to the modern (for example Colley 1992; Nava and O'Shea 1996; Schwarz 1996; Samuel 1998). This volume complements recent historical collections such as Daunton and Reiger's *Meanings of Modernity* (2001), addressing Britain from the late Victorian era to 1939, and Conekin, Mort and Waters' *Moments of Modernity* (1999), concerned with the period 1945 to 1964. The essays in this volume approach the geographies of British modernity through a variety of spatio-temporal themes: longitudinal analyses of social and political change, studies of national identity, archaeologies of particular sites, and discussions of the nature and scale of geographical knowledge. The temporal coverage of individual essays varies: some focus on specific moments in the century, others provide overarching surveys. Here we offer some broad introductory outlines of the geographies of the British modern.

British Modern: Something Done!

British reactions to twentieth-century modernity have been extremely variable. In recent decades a number of commentators have diagnosed a form of British disease – in essence a set of negative responses to the modern world. For example, Martin Wiener (1981) influentially argued that the idealization of the past, and particularly of an aristocratic, deferential, and rural version of the past, was endemic to British culture and had hamstrung economic flexibility and progress in the twentieth century. Such 'declinist' concerns are themselves not new (Friedberg 1988) – indeed one can detect a variant of this theme in the work of Mackinder, whose reaction to early twentieth-century transformations was decidedly anxious and pessimistic. Mackinder's work illustrates the ways in which geographical scholarship has always been a part of wider cultural commentary on the world, a theme to which we return below. In the 1880s, the reaction of the British establishment to modern transformations in space and time had often taken the form of enjoinders to ever greater national effort and enlargement of Britain's global role. J. R. Seeley's 1883 essay on *The Expansion of England* can be read in just this way, not only as a statement of Britain's manifest destiny and its global civilizing mission, but also as a necessary response to the challenges posed by modernity. Similarly, James Froude in his *Oceania or England and her Colonies* of 1886 stressed the need for an ever-extending role in the world:

> The oak tree in the park or forest whose branches are left to it will stand for a thousand years; let the branches be lopped away or torn from it by the wind, it rots at the heart and becomes a pollard interesting only from the comparison of what it once was with what fate or violence has made it. So it is with nations.... A mere manufacturing England, standing stripped and bare in the world's market-place, and caring only to make wares for the world to buy, is already in the pollard state; the glory of it is gone for ever. (1886: 387)

By the beginning of the twentieth century, the response to change had often become distinctly more pessimistic. Mackinder's view of Britain as an 'organic community in decay' in the face of the forces of early twentieth-century modernity was part of a developing tradition that highlighted relative decline – regarding Britain as, in Aaron Freidberg's term, the 'weary titan' (Ó Tuathail 1992: 109; Freidberg 1988). Mackinder, in work from the early 1900s through to the 1940s, provided a conservative geographical analysis that sought to counter a loss of leadership and community with schemes for the maintenance of imperial order through a form of spatial organization that stressed the significance of national, regional and local scale in economic and cultural life. Geographical knowledge was itself a key component of this life:

> It is essential that the ruling citizens of the world wide Empire should be able to visualise distant geographical conditions... Our aim must be to make our whole people think Imperially – think that is to say in spaces that that are worldwide – and to this end our geographical teaching should be directed. (1907: 37 quoted in Ó Tuathail 1992: 114)

While Mackinder offered an anxious and sometimes pessimistic analysis of decline, such a passage also indicates a proponent of what Alison Light has called, in a very different context, 'conservative modernity', characterizing a particular kind of British reaction to substantial social and cultural change: 'Janus-faced it could simultaneously look backwards and forwards; it could accommodate the past in the new forms of the present; it was a deferral of modernity and yet it also demanded a different sort of conservatism from that which had gone before' (1991: 10). We can begin to draw out these concerns via a later specific document of self-consciously British modernity, the 1947 Central Office of Information publication *Something Done: British Achievement 1945–47*. If Mackinder's work gives an academic geographical understanding of Britain, here we find another rendering of Britain's modern space, which is in its own way a geographical account. A heroic steelworker stares from the front cover, a heroic housewife hangs washing on the back (figure 1.1). The publication celebrates achievement in the name of the people during and, despite austerity, carrying wartime rhetoric and publication style into peace. Inside the front cover, over a backdrop of firework celebrations on the Thames in London, Lionel Birch's text begins:

> We came out of the war victorious. We also came out of it much poorer, and needing a rest. There was no rest. There was work to do.
> There were houses and power-stations to build, coal to dig, cloth to weave, ships to launch, fields to till, our trade to rebuild. There were social reforms to make – reforms we all agreed, during the war, that we must have.
> So we came away from the battlefields of the world only to find new and different battles to fight. This book tells the story of some of the first victories. Here are reflected, as in a mirror, certain highlights of achievement – some things we have done as a people, things in which each one of us may take a true national pride.
> In this mirror, and behind these achievements, we see also ourselves – a free people, on the move, in its ancient home. (COI 1947: 2)

The publication begins with a pronatalist celebration of the upturn in the birthrate, linked to 'the problem of maintaining Britain's industrial and cultural potency in a world which is very much on the look-out for any symptoms of British senility' (1947: 6). The back-cover housewife is to produce babies as well as homes (Riley 1981). Industrial and cultural potency are backed up by accounts of power stations, development areas, television, coal, education, new towns, hydro-electric power, aviation, underground railways, films, agriculture, exports, housing, mapping, shipping and the land speed record. Such a publication connects to a wide range of planning and reconstruction literature, and anticipates guides to the Festival of Britain pavilions four years later (Banham and Hillier 1976;

Figure 1.1. Steelworker and Housewife. Front and back cover of *Something Done: British Achievement 1945–47*

Matless 1998). What is striking in this context is the acutely geographical vision of achievement produced. This is in part a matter of demonstrating the physical products of reconstruction, but the new Britain being conjured in these pages in many ways resembles a geographical textbook, with aerial photographs of south Wales industrial estates, Hebburn shipyards, housing estates, London docks, new towns, and airfields turned to arable land. Regional development and global air routes are mapped, while diagrams show HEP stations and tractor production. The spirit of the Ordnance Survey, celebrated for 'making Britain the best-mapped country in the world' (COI 1947: 54), and itself taken as a sign of modern advancement, pervades the document as a whole. Geographical order – modern spaces well laid out, appreciated from the air, integrated into a modern regional organization – is offered as the facilitator and outcome of British achievement. And mapping itself denotes modern life, the account of the Ordnance Survey beginning: 'During the war millions of people in Britain learned for the first time how to orient themselves, and how to find a rendezvous from a map' (1947: 54). Changes in war and peace demand that up-to-date maps are maintained:

> a map is a representation of the ground; and the ground, in this case, is Britain. It is a spacious ground and a varied one, and, since it is not given to any man to go all over the ground before he dies, the next best thing is for him to make an understanding study of the Post-War Ordnance Survey – and to take his choice of Britain. (1947: 55)

Mapping could not only underpin progress but cultivate citizenship (Weight and Beach 1998).

The geography of *Something Done* reflects a fairly conventional mid-twentieth-century sense of a modern planned economy and society, expressing a landscape which it was hoped would further a post-war social democratic consensus of stable and harmonious class relations, advancing the cause of labour through reform rather than revolution. The status of women might also advance without overturning traditional domestic gender relations. Such visions were of course highly contested, but this official document plays down any controversy. Throughout the document material production, modernized through expert knowledge, carries symbolic weight; power generation, new homes, modern mining, ships and steel. This very solid modern geography shaped official senses of mid-twentieth-century Britain and Britishness, just as stories of imperial and globally commercial geographies shaped early twentieth-century accounts (Driver 2001). Different, but no less geographical, stories of a 'late modern' or 'post-modern' economy and society structure accounts of late twentieth-century British modernity. The electronic service economy carries its own symbolic geographies, even in its more extreme rhetorics of footloose life and global interconnection. Declarations that space has been overcome are no less geographical than statements of the value of local rootedness. Zygmunt Bauman has recently described the late twentieth century as an era of 'liquid' or 'fluid' modernity, in contrast with earlier times, characterized by modernity in its 'heavy, bulky, or immobile and rooted solid phase' (Bauman 2000: 57). If, however, the language used to interpret this transformation is explicitly geographical, in such broad analyses there is often neglect of specific national, regional and local formulations. It is instructive to consider late twentieth-century parallels with *Something Done*, official understandings of the nature of Britain and its place in the world. Both New Labour's early attempts to 'rebrand Britain' and the Millennium Dome worked through new types of spatial rhetoric. In *Britain TM*, Mark Leonard's book for the think-tank Demos, Britain's identity could be renewed by understanding the nation as a dynamic and creative 'hub' in a world of networks, flows and connections (Leonard 1997). Similarly, the Greenwich Dome was structured around different zones, many of which attempted to invoke new senses of the 'fluidity' of late twentieth-century Britain: 'journey zone', 'money zone', 'talk zone', 'shared ground'. The perceived failure of the Dome as a national exhibition comes in part from the nature of its times. In previous great exhibitions it was possible to produce state-sanctioned versions of what Dean MacCannell called the 'ethnography of modernity', which drew heavily on exhibiting and staging the fixed sites of work (MacCannell 1999: 13). In the 1951 Festival of Britain, one of the most successful exhibits was a 'working' coalmine on London's South Bank. The Dome suffered from the difficulty of producing equivalent simulacra of spectacular emblematic sites of late twentieth-century British modernity: internet cafes, shiny call centres, expanded universities and science parks failed to capture the imagination in the same way as the power stations, mines and ships of an earlier time. We offer further reflections on such themes in the afterword at the end of this volume.

The official qualities of *Something Done* or the Dome also alert us to the ways in which statements on a country's geography work through claims to cultural authority reflecting particular political projects and economic interests. Lionel Birch's passage quoted above specifies a modern 'we' in whose name *Something Done* speaks and whom it seeks to uphold, a collective 'people' acting together in a common interest. The wartime resonance is obvious, but as many have argued, such collective statements need careful scrutiny, and both the political left and right have then and now questioned the degree to which government could speak as and for the people in the name of reconstruction. The present book seeks to further the scrutiny of geographical knowledge claims, whether official or otherwise, suggesting that it is not enough simply to highlight the importance of the geographies of modernity; we must investigate critically how such geographies have been articulated. The language of geography demands as much critical analysis as any historical narrative. It would, however, be a mistake to argue for geography over history, space over time. This is collection of works in historical geography, and the point throughout is to bring together historical and geographical analysis. Indeed *Something Done* itself requires scrutiny in terms of its historical as well as geographical imagination. At this fulcrum of the twentieth century we find, as Conekin, Mort and Waters (1999) have emphasized, versions of the modern which are adamantly grounded in tradition, seeking to develop rather than replace supposed national qualities and traditions, building a modernism connecting past, present and future rather than signalling temporal and historical rupture. Mid-twentieth-century British arguments around modernity often argued for a blend of the national and international, deploying a historical narrative whereby the Victorian era was set up as a time of industrial and aesthetic chaos, while the Georgian period offered a time of aesthetic order which could be regained and transformed through a new modern spirit. Much the same tactic of historical reclamation and progressive development ran through architectural modernisms of the time (Saint 1987; Whiteley 1995).

Something Done offers achievement going beyond unplanned industrialism via expert leadership, scientific knowledge and planning. This is a different narrative of history and authority from that underpinning visions of British modernity in both the earlier and later twentieth century. If the cultural establishment of the Edwardian era looked back to a globally adventuring Tudor England for imperial legitimation (Colls and Dodd 1986), post-war modernity would later be challenged by a Thatcherite reclamation of Victorian values as dynamic and entrepreneurial rather than ornamentally chaotic. As Donald Horne and others have highlighted, such historical narratives connect to regionally specific visions of national quality, as in Horne's formulation of Northern and Southern British metaphors:

> In the *Northern Metaphor* Britain is pragmatic, empirical, calculating, Puritan, bourgeois, enterprising, adventurous, scientific, serious, and believes in struggle...In the *Southern Metaphor* Britain is romantic, illogical, muddled, divinely lucky, Anglican, aristocratic, traditional, frivolous, and believes in order and tradition...(Horne 1969: 22)

While distinctions between Northern and Southern metaphors can be overplayed, bypassing Eastern and Western, and indeed Midland, metaphors and tending to be reduced in some analyses to progressive northern city versus nostalgic south country, debates on the modernity or otherwise of Britain and its constituent nations and regions always work through senses of history and geography combined at different scales. The story remains to be told of how in the late twentieth century the south-east of England came to epitomize for both proponents and critics a vision of a new country, car-based, consumerist, property-owning, dynamic and/or fraught, continually on the move and looking for settled homes. 'London and its region' becomes once more the centre for national geographical debate, though in a different fashion to late nineteenth-century imperial or interwar/wartime planning argument. This is a region with a complex internal symbolic geography of metropolis, suburb, motorway, countryside, Green Belt and estuary. While recent studies have explored elements of the economic and social geography of 'a neo-liberal heartland' (Allen et al. 1998), a more cultural and historically informed story awaits detailed elaboration. The Southern national metaphor as applied to the south-east breaks down into sub-regions, counties, Downs, corridors, motorway circuits (Brandon and Short 1990). Terms such as Essex and M25 become loaded with symbolic geography.

Critiques of the national geography envisaged in *Something Done* came well before Thatcherism. Senses of post-war Britain as not so much well-ordered as stultifyingly conformist, less progressive than stuck in traditional routine, figure from the late 1950s in the work of diverse authors such as Colin MacInnes, setting a mod London Continental cosmopolitanism against squarely provincial believers in the welfare state (MacInnes 1959), or Keith Waterhouse setting the dreaming Billy Fisher against the honest boredom of provincial family life. The open sequence of John Schlesinger's 1963 film of *Billy Liar* pans along rows of identical flats and suburban semis, while 'Housewives' Choice' plays the request of a lucky housewife (Kenneth McKellar's 'Song of the Clyde'), old urban terraces are demolished and a conventional new order moves along its merry way. National radio picks out individuals who in their choices and houses appear all the same. Upstairs from the family breakfast, lying in late again, Tom Courtenay as Billy gazes at the ceiling and imagines another life: 'Lying in bed, I abandoned the facts again and was back in Ambrosia' (Waterhouse 1959: 5). Schlesinger and Waterhouse tap into mid- to late fifties architectural and social critiques of 'subtopia', instigated by Ian Nairn as an argument against 'making an ideal of suburbia . . . Philosophically, the idealization of the Little Man who lives there' (Nairn 1956: 365). If *Billy Liar* could signal a failed revolt through fantasy, others offered more direct action against a post-war order, whether through radical community politics responding to modernist urban planning, or conservative evocations of local character destroyed by progress. Social, feminist and environmental critiques of the something that had been done since 1945 could take many political forms, which are beyond our discussion here (Roberts 1991; Samuel 1994;

Veldman 1994; Wright 1985, 1991, 1995). Ironically the politically tri-
umphant Thatcherite critique would in its own way idealize that which
figures such as Nairn condemned, with subtopia rescripted in the name of a
suburban individualism devoted to conformities of style, and characteristic-
ally mixing the freedoms of property with devotion to law and order and the
policing of threat and deviance. The landscapes of Thatcherism, as those of
earlier social visions and critiques, carried their own contradictory geographies, which warrant detailed historical exploration.

Our choice of *Something Done* as a specimen of British modernity could
also of course be read as indicating that there is nothing unique to Britain in
such material, as an example of general processes of modernity receiving
British inflection. There are, for example, affinities here with much of the
material covered by Paul Rabinow in his book *French Modern*, where he
elaborates a 'middling modernist' outlook on 'social modernity' over a
hundred years to 1939. Rabinow addresses French variants of international
social scientific ideas and practices, and says of his book that: 'its unit of
analysis is not the nation, the people, or culture, and even less some
perduring "Frenchness"', but is about 'the elements of one specific con-
stellation of thought, action, and passion' (Rabinow 1989: 13). The British
modern addressed in this volume is similarly not explored as part of a
search for an essential Britishness in modernity, or modernness in Britain,
yet that does not mean that one should dismiss the claims to Britishness (or
Englishness, Scottishness, etc.) which may have driven modern British acts
in the twentieth century. Rabinow goes on to argue that: 'An ethnographic
approach to society as the product of historical practices combining truth
and power consists of identifying society as a cultural object, specifying
those authorized to make truth claims about it and those practices and
symbols which localized, regulated, and represented that new reality
spatially' (1989: 13). A key element of those truth contests and spatial
practices has been the way in which senses of the nation have informed
and shaped the processes under scrutiny. To address the British modern is
therefore to consider both how twentieth-century Britain carried a specific
and unique mix of global and local economic, political, cultural processes,
and how those processes were inflected by a sense of Britain as modern or
otherwise.

British Historical Geography and the Twentieth Century

This collection presents work in historical geography. As we have argued
above, the most important justification for this is the inherently spatialized
character of both twentieth-century modernity in general and the British
experience in particular. There is also a significant secondary motivation for

this collection, in that it reflects a developing focus in a sub-discipline. The past 20 years have seen a substantial shift in the temporal balance of studies in historical geography, with much more attention being paid first to the inter-war period (Heffernan and Gruffudd 1988), and then increasingly to the years after the Second World War. To some extent this is a simple consequence of the passing of time, as events, processes and patterns from the twentieth century have come to be defined as being of 'historical' interest, much as 'contemporary' history has emerged in recent decades. Thus the major existing collection on the historical geography of England and Wales, itself concerned for earlier geographies of modernity, stops at 1900 (Dodgshon and Butlin 1990). The interest in new time periods also reflects, then, a growing interest in the sub-discipline in the conceptual challenges posed by the distinctive formations of twentieth-century modernities.

The twentieth century is distinctive not least as the period when geography as a new academic and institutionalized practice itself played a role in the processes under scrutiny, reflecting and forming perspectives on the world, whether through the observation or education of those concerned with the planning of industrial change, the governance of empire, or the contestation of cultural identity. The twentieth century is special for us in part because it contains the genesis and development of British geography at university level. Historical geography developed as a sub-discipline distinct from both history, with which it had been so closely intertwined in the writings of Paul Vidal de la Blache (1843–1918) in France, and anthropology, where links had been personified in H. J. Fleure, Professor of Geography and Anthropology at Aberystwyth (Campbell 1972; de Planhol 1972). The writings of earlier generations of contemporary human and environmental geographers are available to us now as representations of earlier geographies, and they appear sometimes as sources, but more often as objects of analysis and interpretation.

There are many overviews of the sub-discipline available which chart its development through the century (Baker and Billinge 1982; Butlin 1993; Darby 1983, 1987), and we do not propose to conduct a further full review here. However, it is important to note that historical geography has repeatedly found itself a testing ground for theory as paradigms and intellectual fashions have changed, a springboard for methodological pluralism, and a source for empirical data collection. As Harley somewhat caustically put it: 'Theories, ill-clad and poorly shod with evidence, migrated like paupers into the past, lured on by the prospect of factual El Dorados to validate universal truths' (1982: 264). An element of reflexivity becomes almost inescapable, as the development of historical geography has both reflected and informed perspectives on the twentieth-century world.

The term 'historical geography' was well established by the beginning of the twentieth century, when it referred to attempts to demonstrate the significance of geographical features in the study of a more realistic history, and more broadly (and deterministically) to refer to the study of the influence of geography upon history. In the mid-twentieth century the most important figure shaping the sub-discipline in Britain was probably H. C. Darby. Particularly through his role as professor at Cambridge University, Darby was an important influence on succeeding generations of historical geographers – as both inspiration and increasingly as the focus of critiques (Prince 2002). In the late 1920s Darby's first exposure to Cambridge historical geography was through Bernard Manning's final-year course which focused on the geographical conditions affecting the historical and political development of states, and which also took in colonial expansion and political subdivisions for administrative purposes (Darby 1987: 117–37, 2002: 1–2). This close relationship with political and colonial issues had been typical of the sub-discipline during what Butlin (1993: 3) refers to as the period of 'Nationalism 1870–1914', typified perhaps by the Revd H. B. George's *Historical Geography of the British Empire*, first published in 1904 and reaching a seventh edition by 1924.

However, significant changes took place from the 1920s onwards, stimulated in part by the establishment and growth of geography departments in many British universities. The concern with explicitly political and colonial issues was downgraded as Darby, with others, moved the focus to an intensely academic and empirical concern with work on the Fens, on the Domesday Book, and on woodland clearance. This work applied what were self-consciously understood as 'modern' geographical methods to historical data, drawing upon contemporary developments in regional and environmental analysis (Baker 1972, 2003; Butlin 1993). Darby became increasingly concerned to go beyond the detailed cross-sectional reconstructions of past landscapes at particular moments that were seen to best effect in the Fenland work or in his edited 1936 *Historical Geography of England before 1800*. Later work attempted to combine the established cross-sectional method with concerns for process and narrative to incorporate the dimension of change. The updated 1973 volume, the *New Historical Geography of England*, shows the full extent of the development in Darby's vision of the sub-discipline, with cross-sectional essays interleaved with discussions of process.

Darby's work on the Fens had emphasized a heroic, even Whiggish drive for the taming of nature, echoing other visions of British progress and position in the world. Darby himself was part of a certain culture of mid-century British progressivism, particularly through his public service as a national parks commissioner after the war (Darby 1961). In such work a vision of a new order of popular citizenship with access to the countryside was combined with a concern for the conservation of historic landscapes. Darby's supposed academic neutrality and progressive vision were, as he later acknowledged, 'prisoners of their own time and of their own cultural and intellectual world' (Darby 1983: 423). In retrospect, the *New Historical Geography of England* seems like the last significant example of this way of presenting historical geography. Tellingly the final cross-sectional essay, by Peter Hall, concerns 'England circa 1900', almost as if the challenges of sustaining the method into the period of living memory were too great. As Darby himself recognized obliquely in his introduction to the volume, from the 1960s onwards the impact of wider changes within geography had ensured that the sub-discipline would change significantly:

> In another generation or so the materials for an historical geography of England will not be as we know them now. A wider range of sources will have been explored and evaluated. Fresh ideas about method will have prepared the way for a more sophisticated presentation. And by that time we and our landscape will have become yet one more chapter in some other *Historical Geography of England.* (Darby 1973: xiv)

What for Darby could be portrayed as the increasing sophistication of new methods was in fact a series of quite fundamental challenges to the established approach of historical geography – what Butlin has described as its 'classical phase' between 1930 and 1960. From the 1960s onwards, as is well attested, geography moved in concert with other disciplines to successively embrace the several statistical, theoretical and behavioural approaches, and these found their place within the sub-discipline with varying degrees of success and contestation. Regional studies were joined by work on perception and cognition by Prince (1971: 1–86), Lowenthal and Bowden (1975) or Powell (1977), who thereby again imparted historical concerns to the prevailing geography of the day. A more sustained critique of established historical geography came from those geographers influenced by Marxism, who worked on the historical materiality of space (indeed, pushing space into historical materialism). From this perspective the progressive narrative implicit within Darby's historical geography could be regarded as 'materialistic and bourgeois' (Gregory 1984: 186), and as over-consensual, partial and even elitist. Criticisms of 'the Darbyesque landscape' saw it as largely devoid of human agency: 'those appearing were the rulers and directors only' (Williams 2002: 207–8). The work of David Harvey in particular prompted a shift towards a focus on the historical geography of capitalist development and of class conflict (Harvey 1973, 1979), and Harvey's theoretical work was joined by the more empirically informed historical depth of Massey's spatial divisions of labour within

Britain (Massey 1984), presenting rounds of successive regional investment as forming the creative destruction of capitalist modernity. In the 1980s, working within a broad cultural Marxist framework, Cosgrove (1984) and Cosgrove and Daniels (1988) revealed the cultural and power-laden asymmetries of the landscape, one of the central and relatively unproblematic themes of the older historical geography. The differences between the first and second editions of Dodgshon and Butlin's *Historical Geography of England and Wales* indeed reflect such changes within the sub-discipline between 1978 and 1990.

One feature of this shift towards historical geographies of capitalist development and social and cultural conflict was the way in which it opened new possibilities for the study of the twentieth century. The experience of restructuring and of renewed mass unemployment in the 1980s stimulated a new generation of historical geographers to examine the inter-war period, a time that seemed to offer relevant evidence about the role of capital and state in the development of regional inequalities, the struggles for survival of particular places in the face of powerful economic and political forces, and the organization of collective resistance, particularly through the union movement (Ward, 1988; Heffernan and Gruffudd 1988; Sunley 1990; Gilbert 1992). The cross-sectional genre was transformed into studies of different aspects of industrialization and its consequences (Langton and Morris 1986), or of social and political conflict (Charlesworth et al. 1996). In keeping with wider cultural and political agendas, the locus of such political enquiry went beyond workplace studies of class relations to encompass matters of gender and community, as in Rose's work on inter-war east London (Rose 1990, 1997). Attention to the economic and political dimensions of the experience of British modernity was here accompanied by work that sought to broaden this understanding, again often revisiting established sub-disciplinary themes from perspectives that drew upon critical traditions interrogating the nature of modernity. A number of studies addressed relations between city, country and suburbia, and the ways in which each can become the focus of particular cultural values concerning landscape, community and place. Such work brings out in different contexts the geographies of power and identity associated with rurality, urbanity and the suburban, and connects also to an interest in relations of landscape and national identity. (Matless 1998; Short et al. 2000) Work in the late twentieth century has been characterized by a renewed interest in the distinctiveness of such relationships in the constituent nations of the United Kingdom (Gruffudd et al. 2000; Lorimer 2000, 2001), work which brings out the complex relationships between Britishness and Englishness, Welshness and Scottishness, notably in terms of resistance to one in the name of the other (Gruffudd 1995, 1999). The work of Nash and Graham on Irish landscape, like Kearns's essay in this volume, addresses a very different politics of culture and landscape in relation to a British presence and/or legacy (Graham 1997; Nash 1996, 1999).

Other recent studies draw out the geographies of British modernity and Britishness in more direct connection to Britain's global imperial role, and its legacies across the world and at home, in studies of the historical geographies of colonialism and post-colonialism (Lester 2000). The British empire is also the subject of work on the geographies of migrant domestic identity in British women of empire, the photographic geographies of empire, the remaking of Britishness itself through processes of post-colonialism, and the imperial urbanity of early twentieth-century Britain, whether articulated in buildings, travel guides, exhibitions or pageants (Blunt 1999; Blunt and Rose 1994; Ryan 1997; Schwarz 1996; Driver and Gilbert 1999; Gilbert 1999). Such work underlines the point that geographies of Britain and Britishness can never be understood solely within the formal cartographic outline of Britain, but rather should attend to both internal heterogeneity and relations with elsewhere. The latter is of course not only connected to empire and the post-colonial nature of contemporary Britain. Europe and the USA provide touchstones for Britishness and for Britain's global role throughout the twentieth century, whether in battles for global trade and influence, cultural pro- or anti-Americanism, or admiration or disdain for the culture and polity of Europe. Such issues of internationally comparative and connective historical geography represent a neglected area demanding further investigation (Graham 1998; Heffernan 1998).

In recent reflections on Darby's work Michael Williams (2002) has argued that the traditional concerns of historical geography risk being swallowed up by the wider cultural turn in geographical enquiry on the one hand, and the more specific presence of environmental history on the other (cf. Simmons 1993, 1998). In part though the modern distinctiveness of British historical geography, compared for example with that of the USA, continues to rest in an appreciation of the material and ideological landscape as a cultural product, compared with environmental histories of the North American landscape which have worked predominantly through a narrative of 'nature' as wilderness eroded or destroyed. In this sense relations of historical geography and environmental history have a geography of their own. The relative decline of interest in environmental issues within British historical geography as a result of academic fragmentation and specialization does, however, raise another issue, namely the extent to which such specialization reduces the integrative power of historical geography. Synthesis may be sought, but the time/space/theme specialization now required of research students and by those institutions and procedures policing output, such as the Research Assessment Exercise, precludes many geographers from tackling fundamental ecological or spatial issues within historical geography. It appears to be something of a paradox that while the range of materials tackled, methods employed, interdisciplinary links exploited – with sociology, political economy and anthropology, for example – all increased greatly during the second half of the twentieth century, there are some respects in which historical geography has narrowed. This volume cannot be totally exempt from such a charge. Like early twenty-first-century historical geography more generally, it concentrates particularly on the socio-cultural historical analysis of twentieth-century Britain, rather than on, say, environmental concerns. Across the later twentieth century, geography as a discipline witnessed the reconfiguration of the economic, the cultural, and the social within human geography. This is not to suggest, for example, that the 'cultural' triumphed over the economic, but to agree with others from within geography and beyond (Geertz 1983; McDowell 1994) that there has been a blurring of genres, a convergence of attitudes and methodologies that has obviated the previous schisms. Examples can be found of this 'cultural turn' within the chapters of this book, and one could indeed argue that it was within historical geography (broadly defined) that such reconfigurations have been most evident, indeed were pioneered.

In 1972 Baker voiced the need to 'face squarely the current intellectual crisis' caused by the rift between historical geography and geography as the latter had moved (temporarily) into its generalized model-possessed phase (1972: 12). By 1988, however, Driver could claim with conviction that 'any division between a non-historical human geography, oriented to the present, and an historical geography oriented to the past can no longer be sustained' (1988: 504). The boundaries between historical geography and other branches of the discipline have become more permeable as the necessity for investigating cultural, social or economic depth of process or agency has been recognized (see Dennis 1989 on urban geography, or Dunford and Perrons 1983 on economic change). Human geography is increasingly configured in ways which appreciate the insights to be gained from geographies which are viewed historically (Gregory, Martin and Smith 1994: 1–11). Despite the unease amongst some historical geographers that the sub-discipline might thereby be assimilated into a wider geographical project and lose its distinctive insights (Butlin 1993: 45), today the two are much closer intellectually, to the advantage of both, as is demonstrated by the essays in this collection. In similar fashion historical geographers have been happy to embrace connections with re-visioned historical projects of spatialized or geographical history. Indeed Philo, writing on historical

geography in 1994, suggested that 'study after study now actually proceeds in the vein of a geographical history where the focus has shifted from the materiality of "geographical facts" to the immateriality of historical phenomena... much research has turned to phenomena with at best a minor or tangential impact upon "the soil"' (Philo 1994: 259). Darby was unsympathetic to such a development, with its overtly social science origins, and which chose to focus on people rather than places, and people as changers of places rather than on places transformed by people (Darby 1983: 421–8, 2002: 25). Philo's comments anticipate Elden's call to spatialize history, by injecting an awareness of space and the ways 'space is effected and effects' (Elden 2001: 7), and this collection proceeds in a similar way. The volume sits, however, somewhere between Philo and Darby's statements, with some chapters beginning from social, economic or political issues to examine their spatial expression, and others using sites and places, soil-based or otherwise, as starting points for their social enquiries.

A final point on the historical geography of the twentieth century concerns source material. Research both draws on new sources, and encounters limitations – both legal and ethical – placed on the use or reproduction of such sources. The writings of twentieth-century geographers and others have in themselves created a huge reservoir of material, and within geography new sources/representations have been studied (Short 1997; Short et al. 2000). If, however, much statistical data is available at national or sub-national levels, many near-contemporary primary sources at individual or community levels are not yet available. At the time of writing, for example, only the 1901 census has been put online from the Public Record Office, where various 35-, 50- and 75-year closure periods necessarily restrict access to modern sources for the geographer in ways that no longer affect those working, for example, on Victorian Britain. This fact has lent weight to those historical geographers who have developed research links with other disciplines in order to develop new tools and agendas for working with the modern subject. We can no longer hide behind a comfortable separatist view that historical geography is based on 'enthusiasms and insights generated by its own practitioners without drawing much on concepts and methods developed in other fields' (Baker 1982: 234). Recent work has drawn upon methods such as oral history, textual analysis, biography and ethnography; the new media of the twentieth century – radio, cinema, television – and the records they have left also offer important potential sources, as yet under-used in the sub-discipline. The essays in this collection are formed from a wide range of material – official statistics, qualitative surveys, popular literatures, commercial journals, political tracts, oral histories, built environments, academic geographies – and signify a historical geography enriched in both methods and sources.

Approaching the Geographies of British Modernity

We have structured this volume in three sections, around themes of survey, site and identity. The essays have been grouped according to the approach through which they explore their particular materials, and indeed the three sections correspond to three modes of geographical thinking and modern sensibility. First we have a section informed by the survey motif common in twentieth-century thinking, whereby modernity is claimed for a way of thinking which seeks a rationally ordered overview of particular situations. While the essays here employ different sources and methods, from statistics to qualitative social surveys, all seek a general perspective on a key geographical issue; poverty, voting, commuting, rural change. Dorling draws and reflects upon the tradition of social statistics to outline the geographies of inequality through the twentieth century, detailing relative and absolute issues of continuity and change over a hundred years. Johnston and colleagues offer a survey of electoral geography through the century, asking whether and how the period can be characterized as a 'Conservative century'. Again surveys are scrutinized as they are utilized, with issues of electoral bias and gerrymandering the focus of analysis. Pooley draws on a major study of population mobility in the twentieth century to address specific issues of journey to work in relation to residential relocation. Broad statistical coverage is integrated with oral histories to draw out individual experiences within aggregate trends. Howkins further develops qualitative as well as quantitative surveys by utilizing the results of a recent Mass Observation Directive concerning perceptions of change in the British countryside since 1945. Issues of agricultural transformation and environmental change are drawn out from the testimony of individual respondents, and connected to ongoing debates on the future of farming and landscape.

The second group of essays seeks general understanding through the particular, taking specific sites, whether individual localities or types of place, as ways into general matters of social, cultural and economic life. The motorway, the industrial estate, the shop and the mosque become sites through which to approach the British modern. Merriman explores the building of the M1 as an occasion for various articulations of modernity, whether in construction and landscaping projects or cultures of driving. A range of official and unofficial sources are drawn upon to make up a rich geography of a modern road. Linehan similarly explores a vision of modern landscape through inter-war industrial spaces, outlining general discourses of industrial survey and plan before considering in detail the Team Valley estate in Gateshead. The seemingly prosaic spaces of the industrial estate become sites to envisage a new England, with political, regional, commercial and expert interests combining to produce modern spaces. Edwards draws out a different space of mid-twentieth-century modernity through Simpson of Piccadilly, a commercial site fostering a mode of modern consumption rather than production. Modern British, and specifically English, masculinity is produced through a self-consciously modern yet traditional commercial space in London's West End. Naylor and Ryan consider non-Christian spaces of worship in twentieth century British cities, spaces less self-consciously modern yet understandable only through processes of British modernity. Imperialism, post-colonialism and suburbanization come together in the mosques, gurdwaras and mandirs developed as part of the reconfiguration of the British religious landscape over the past hundred years.

The third section addresses the specific theme of relations of geography, nation and identity, with essays taking very different routes into the spaces of national identity produced through domestic, local, national, imperial and post-imperial commentary. Gilbert and Preston further explore the landscapes of English suburbia, highlighting their twentieth-century incorporation into cultural and political narratives of Englishness. A form of historical political geography contrasting to that in Johnston's essay draws attention to the symbolic geographies of political leadership, with the embrace or rejection of suburban qualities by political leaders. Other landscapes of political identity are addressed by Kearns in his analysis of the geographies of early twentieth-century Irish nationalism. Kearns highlights the different geographies informing variants of nationalism, and characterizes nationalism as an adamant political discourse nevertheless characterized by an ambivalent politics of scale, with the national, the local, the universal and the cosmopolitan operating in uneasy relationship. Butlin brings the issue of academic geographical knowledge and its wider political role to the fore, discussing the ways in which forms of geography bound up with British state policy overseas developed in relation to continuing projects of empire and later processes of decolonization. At times geography as practised through the Royal Geographical Society has been characterized by forms of political reticence, at others by a sense of academia on active service.

While only Butlin's essay here considers specifically academic geography, the theme of the production of geographical knowledges runs throughout the volume. In 1986 Peter Taylor offered a fourfold typology of geographical knowledge: the 'necessary' knowledge used in everyday life, the 'professional' knowledge of academic geographers, the 'gainful' knowledge of commercial and state intelligence, and the 'popular' knowledge of travellers' tales and guides. Essays in this volume present the works of geographers, political activists, cartographers, industrialists, travellers and others as forms of modern geographical knowledge reflecting and contributing to the geographies of British modernity. If the volume overall is an exercise in the historical geography of twentieth-century Britain, its subject matter is in large part made from forms of geographical knowledge circulating through that time and place.

The themes of survey, site and identity offered in this volume complement or build upon the work by historical geographers identified above to address some of the key issues confronting Britain in the distinctive historical period of the last century. The chapters shed new light on the social atomization and persistent inequalities within British society; the wrenching apart of the British nation-state; and on decolonization. There are, of course, many other vital themes which we acknowledge but which cannot fully and explicitly find a place here: the fundamental changes in the spatial pattern of British external relations, especially with Europe; the two catastrophic wars from which Britain emerged in poverty; deindustrialization and its concomitant ongoing scientific and technological revolutions, the geographies of changing gender roles, organized labour, civil rights movements, immigration and ethnic pluralities, and the paradox of increasing penetration by the state into everyday life at a time when the state itself was under pressure from supra-national regionalism and globalization. At particular historical moments all such issues clearly helped shape the geographies of British modernity. Such changes transformed many of the things which at the start of the century had constituted a sense of 'Britishness' and they certainly called for adjusted ideas about Britain's place in the world.

Looking back over the 'tectonic upheavals' of the twentieth century, Hobsbawm reflected in geographical vein on 'the American century' which had not ended well for Britain:

> It has for the first time become possible to see what a world may be like in which the past, including the past in the present, has lost its role, in which the old maps and charts which guided human beings, singly and collectively, through life no longer represent the landscape through which we move, the sea on which we sail. (Hobsbawm 1994: 16)

The temporal schisms of the century have been explored elsewhere – this volume is a contribution to understanding the processes, outcomes and impacts of the spatial schisms within twentieth-century Britain which, as we suggest in our Afterword, have a continuing pertinence in the Britain of the twenty-first century.

REFERENCES

Allen, J., D. Massey and A. Cochrane 1998: *Rethinking the Region*. London: Routledge.

Baker, A. R. H. 1972: Rethinking historical geography. In A. R. H. Baker (ed.), *Progress in Historical Geography*. Newton Abbot: David & Charles.

Baker, A. R. H. 1982: On ideology and historical geography. In A. R. H. Baker and M. Billinge (eds), *Period and Place: Research Methods in Historical Geography*. Cambridge: Cambridge University Press, 233–43.

Baker, A. R. H. 2003 (forthcoming): *History and Geography*. Cambridge: Cambridge University Press.

Baker, A. R. H., and M. Billinge (eds) 1982: *Period and Place: Research Methods in Historical Geography*. Cambridge: Cambridge University Press.

Banham, M. and B. Hillier 1976: *A Tonic to the Nation: The Festival of Britain 1951*. London: Thames & Hudson.

Bauman, Z. 2000: *Liquid Modernity*. Cambridge: Polity.

Berman, M. 1983: *All That Is Solid Melts Into Air*. London: Verso.

Blunt, A. 1999: Imperial geographies of home: British women in India, 1886–1925. *Transactions Institute of British Geographers*, 24, 421–40.

Blunt, A., and G. Rose (eds) 1994: *Writing Women and Space: Colonial and Post-colonial Geographies*. New York: Guilford.

Bowden, M. J. 1969: The perception of the western interior of the United States, 1800–1870: A problem in historical geography. *Proceedings, Association of American Geographers*, 1, 16–21.

Brandon, P., and B. Short 1990: *The South East from AD 1000*. London: Longman.

Butlin, R. A. 1993: *Historical Geography: Through the Gates of Space and Time*. London: Arnold.

Campbell, J. 1972: *Some Sources of the Humanism of H. J. Fleure*. Oxford University School of Geography Research Paper no. 2.

Central Office of Information 1947: *Something Done: British Achievement 1945–47*. London: HMSO.

Charlesworth, A., D. Gilbert, A. Randall, H. Southall and C. Wrigley 1996: *An Atlas of Industrial Protest in Britain, 1750–1990*. Basingstoke: Macmillan.

Colley, L. 1992: *Britons: Forging the Nation 1707–1837*. New Haven: Yale University Press.

Colls, R., and P. Dodd (eds) 1986: *Englishness: Politics and Culture*. London: Croom Helm.

Conekin, B., F. Mort and C. Waters (eds) 1999: *Moments of Modernity: Reconstructing Britain 1945–1964*. London: Rivers Oram Press.

Cosgrove, D. 1984: *Social Formation and Symbolic Landscape*. London: Croom Helm.

Cosgrove, D., and S. Daniels (eds) 1988: *The Iconography of Landscape*. Cambridge: Cambridge University Press.

Darby, H. C. 1936: *An Historical Geography of England before 1800*. Cambridge: Cambridge University Press.

Darby, H. C. 1961: National Parks in England and Wales. In H. Jarrett (ed.), *Comparisons in Resource Management*. Baltimore: Johns Hopkins University Press, 8–34.

Darby, H. C. 1973: *A New Historical Geography of England*. Cambridge: Cambridge University Press.

Darby, H. C. 1983: Historical geography in Britain, 1920–80: continuity and change. *Transactions Institute British Geographers*, NS 8, 421–8.

Darby, H. C. 1987: On the writing of historical geography 1918–1945. In R. W. Steel, *British Geography* 1918–1945. Cambridge: Cambridge University Press, 117–37.

Darby, H. C. 2002: *The Relations of History and Geography: Studies in England, France and the United States*, ed. M. Williams, H. Clout, T. Coppock and H. Prince. Exeter: University of Exeter Press.

Daunton, M., and B. Rieger (eds) 2001: *Meanings of Modernity: Britain from the Late-Victorian Era to World War II*. Oxford: Berg.

de Planhol, X. 1972: Historical geography in France. In A. Baker (ed.), *Progress in Historical Geography*. Newton Abbot: David & Charles, 29–44.

Dennis, R. 1989: Dismantling the barriers: past and present in urban Britain. In D. Gregory and R. Walford (eds.), *Horizons in Human Geography*. London: Macmillan, 194–216.

Dodgshon, R. A., and R. A. Butlin (eds) 1990 [1978]: *An Historical Geography of England and Wales*, 2nd edn. London: Academic Press.

Driver, F. 1988: The historicity of human geography. *Progress in Human Geography*, 12, 497–506.

Driver, F. 2001: *Geography Militant: Cultures of Exploration and Empire*. Oxford: Blackwell.

Driver, F., and D. Gilbert (eds) 1999: *Imperial Cities: Landscape, Display and Identity*. Manchester: Manchester University Press.

Dunford, M., and D. Perrons 1983: *The Arena of Capital*. London: Macmillan.

Earle, C. 1992: *Geographical Inquiry and American Historical Problems*. Stanford: Stanford University Press.

Elden, S. 2001: *Mapping the Present: Heidegger, Foucault and the Project of a Spatial History*. New York: Continuum.

Friedberg, A. 1988: *The Weary Titan: Britain and the Experience of Relative Decline 1895–1905*. Princeton: Princeton University Press.

Froude, J. 1886: *Oceania or England and her Colonies*. London: Longman, Green & Co.

Geertz, C. 1983: *Local Knowledge: Further Essays in Interpretive Anthropology* New York: Basic Books.

Gilbert, D. 1992: *Class, Community and Collective Action: Social Change in Two British Coalfields 1850–1926*. Oxford: Clarendon Press.

Gilbert, D. 1999: 'London in all its glory – or how to enjoy London': guide-book representations of Imperial London. *Journal of Historical Geography*, 25, 279–97.

Graham, B. (ed.) 1997: *In Search of Ireland: A Cultural Geography*. London: Routledge.

Graham, B. (ed.) 1998: *Modern Europe: Place, Culture, Identity*. London: Arnold.

Graham, B., and C. Nash (eds) 2000: *Modern Historical Geographies*. Harlow: Prentice Hall.

Gregory, D. 1982: *Regional Transformation and Industrial Revolution: A Geography of the West Yorkshire Woollen Industry*. London: Macmillan.

Gregory, D. 1984: Some *terrae incognitae*. In A. R. H. Baker and D. Gregory (eds), *Explorations in Historical Geography*. Cambridge: Cambridge University Press.

Gregory, D. 1994: *Geographical Imaginations*. Oxford: Blackwell.

Gregory, D., R. Martin and G. Smith (eds) 1994: *Human Geography: Society, Space and Social Science*. London: Macmillan.

Gruffudd, P. 1995: Remaking Wales: nation-building and the geographical imagination, 1925–50. *Political Geography*, 14(3), 219–40.

Gruffudd, P. 1999: Prospects of Wales: contested geographical imaginations. In R. Fevre and A. Thompson (eds), *Nation, Identity and Social Theory: Perspectives from Wales*. Cardiff: University of Wales Press, 149–67.

Gruffudd, P., D. T. Herbert and A. Piccini 2000: In search of Wales: travel writing and narratives of difference, 1918–50. *Journal of Historical Geography*, 26(4), 589–604.

Harley, J. B. 1982: Historical geography and its evidence: reflections on modelling sources. In A. R. H. Baker and M. Billinge (eds), *Period and Place: Research Methods in Historical Geography*. Cambridge: Cambridge University Press, 261–73.

Harvey, D. 1973: *Social Justice and the City*. London: Edward Arnold.

Harvey, D. 1979: Monument and myth. *Annals of the Association of American Geographers*, 69, 362–81.

Heffernan, M. 1998: *The Meaning of Europe: Geography and Geopolitics*. London: Edward Arnold.

Heffernan M., and P. Gruffudd 1988: *A Land Fit For Heroes: Essays in the Human Geography of Inter-War Britain*. Loughborough: Department of Geography.

Hobsbawm, E. 1994: *Age of Extremes: The Short Twentieth Century 1914–1991*. London: Michael Joseph.

Horne, D. 1969: *God is an Englishman*. Harmondsworth: Penguin.

Kern, S. 1983: *The Culture of Time and Space, 1880–1918*. Cambridge, MA: Harvard University Press.

Langton, J. and R. Morris 1986: *Atlas of Industrializing Britain, 1780–1914*. London: Methuen

Leonard, M. 1997: *Britain TM: Renewing our Identity*. London: Demos.

Lester, A. 2000: Historical geographies of imperialism. In B. Graham and C. Nash (eds), *Modern Historical Geographies*. London: Addison Wesley Longman, 100–20.

Light, A. 1991: *Forever England: Femininity, Literature and Conservatism between the Wars*. London: Routledge.

Lorimer, H. 2000: Guns, game and the grandee: the cultural politics of deerstalking in the Scottish Highlands. *Ecumene*, 7, 403–31.

Lorimer, H. 2001: Sites of authenticity: Scotland's new parliament and official representations of the nation. In D. Harvey, R. Jones, N. McInroy and C. Milligan (eds), *Celtic Geographies: Landscape, Culture and Identity*. London: Routledge, 91–108.

Lowenthal, D., and M. J. Bowden (eds) 1975: *Geographies of the Mind: Essays in Historical Geosophy in Honour of John Kirkland Wright*. New York: Oxford University Press.

MacCannell, D. 1999: *The Tourist: A New Theory of the Leisure Class*. Berkeley: University of California Press.

MacInnes, C. 1959: *Absolute Beginners*. London: MacGibbon & Kee.

Mackinder, H. 1902: *Britain and the British Seas*. London: Heinemann.

Mackinder, H. 1904: The geographical pivot of history. *Geographical Journal*, 13, 421–37.

Mackinder, H. 1907: On thinking imperially. In M. Sadler (ed.), *Lectures on Empire*. London: printed privately.

Massey, D. 1984: *Spatial Divisions of Labour*. London: Macmillan.

Matless, D. 1998: *Landscape and Englishness*. London: Reaktion.

McDowell, L. 1994: The transformation of cultural geography. In D. Gregory, R. Martin and G. Smith (eds), *Human Geography: Society, Space and Social Science*. London: Macmillan, 146–73.

Nairn, I. 1956: *Outrage*. London: Architectural Press.

Nash, C. 1996: Men again: Irish masculinity, nature, and nationhood in the early twentieth century. *Ecumene*, 3, 427–53.

Nash, C. 1999: Irish placenames: post-colonial locations. *Transactions Institute of British Geographers*, 24(4), 457–80.

Nava, M., and A. O'Shea (eds) 1996: *Modern Times: Reflections on a Century of English Modernity*. London: Routledge.

Ó Tuathail, G. 1992: Putting Mackinder in his place. *Political Geography*, 11, 100–18.

Ogborn, M. 1998: *Spaces of Modernity: London's Geographies 1680–1780*. London: Guilford

Philo, C. 1994: History, geography and the 'still greater mystery' of historical geography. In D. Gregory, R. Martin, and G. Smith (eds), *Human Geography: Society, Space and Social Science*. London: Macmillan, 252–81.

Powell, J. M. 1977: *Mirrors of the New World. Images and Image-Makers in the Settlement Process*. Folkestone: Dawson, Archon Books.

Pred, A. 1986: *Place, Practice and Structure: Social and Spatial Transformation in Southern Sweden: 1750–1850*. Cambridge: Polity.

Pred, A., and M. Watts 1992: *Reworking Modernity: Capitalisms and Symbolic Discontent*. New Brunswick: Rutgers University Press.

Prince, H. 1971: Real, imagined and abstract worlds of the past. *Progress in Human Geography*, 3, 1–86.

Prince, H. 2002: H. C. Darby and the historical geography of England. In H. Darby, *The Relations of History and Geography*. Exeter: University of Exeter Press, 63–88.

Rabinow, P. 1989: *French Modern: Norms and Forms of the Social Environment*. Cambridge: MIT Press.

Riley, D. 1981: 'The free mothers': Pronatalism and working women in industry at the end of the last war in Britain. *History Workshop Journal*, 11, 58–118.

Roberts, M. 1991: *Living in a Man-Made World: Gender Assumptions in Modern Housing Design*. London: Routledge.

Rose, G. 1990: Imagining Poplar in the 1920s: contested concepts of community. *Journal of Historical Geography*, 16, 425–37.

Rose, G. 1997: Engendering the slum: photography in East London in the 1930s. *Gender, Place and Culture*, 4, 277–300.

Ryan, J. 1997: *Picturing Empire: Photography and the Visualisation of the British Empire*. London: Reaktion Books.

Saint, A. 1987: *Towards a Social Architecture: The Role of School-Building in Post-War England*. New Haven: Yale University Press.

Samuel, R. 1994: *Theatres of Memory*. London: Verso.

Samuel, R. 1998: *Island Stories: Unravelling Britain*. London: Verso.

Schwarz, B. (ed.) 1996: *The Expansion of England: Race, Ethnicity and Cultural History*. London: Routledge.

Seeley, J. 1883: *The Expansion of England: Two Courses of Lectures*. London: Macmillan

Short, B. 1997: *Land and Society in Edwardian Britain*. Cambridge: Cambridge University Press.

Short, B., C. Watkins, W. Foot, and P. Kinsman 2000: *The National Farm Survey 1941–43: State Surveillance and the Countryside in England and Wales in the Second World War*. Wallingford: CAB International.

Simmons, I. G. 1993: *Environmental History*. Oxford: Oxford University Press.

Simmons, I. G. 1998: Towards an environmental history of Europe. In R. A. Butlin and R. A. Dodgshon (eds), *An Historical Geography of Europe*. Oxford: Oxford University Press, 335–61.

Steel, R. W. 1987: The beginning and the end. In R.W. Steel (ed.), *British Geography 1918–1945*. Cambridge: Cambridge University Press, 1–8.

Sunley, P. 1990: Striking parallels: a comparison of the geographies of the 1926 and 1984–85 coal-mining disputes. *Environment and Planning D: Society & Space*, 8, 35–52.

Taylor, P. 1986: Locating the question of unity. *Transactions Institute of British Geographers*, 11, 443–8.

Veldman, M. 1994: *Fantasy, the Bomb, and the Greening of Britain: Romantic Protest, 1945–1980*. Cambridge: Cambridge University Press.

Ward, S. 1988: *The Geography of Interwar Britain: The State and Uneven Development*. London: Routledge.

Waterhouse, K. 1959: *Billy Liar*. London: Michael Joseph.

Weight, R., and A. Beach 1998: *The Right to Belong: Citizenship and National Identity in Britain, 1930–1960*. London: IB Tauris.

Whiteley, N. 1995: Modern architecture, heritage and Englishness. *Architectural History*, 38, 220–37.

Wiener, M. 1981: *English Culture and the Decline of the Industrial Spirit 1850–1980*. Cambridge: Cambridge University Press.

Williams, M. 2002: Epilogue: critique and evaluation. In H. C. Darby, *The Relations of History and Geography*. Exeter: Exeter University Press, 203–11.

Wright, P. 1985: *On Living in an Old Country: The National Past in Contemporary Britain*. London: Verso.

Wright, P. 1991: *A Journey through Ruins*. London: Radius.

Wright, P. 1995: *The Village that Died for England*. London: Jonathan Cape.

Part I

Surveying British Modernity

Chapter Two

A Century of Progress? Inequalities in British Society, 1901–2000

Danny Dorling

In this chapter I argue that people in Britain experienced incredible social progress between 1901 and 2000: a kind of progress that could not have been envisaged at the start of the century. However, at the same time there was far less progress achieved in reducing social inequalities over space and across society. The first half of this chapter looks at absolute progress in British living standards – often overlooked in accounts of British social change. This is particularly true for the last 25 years, when average standards of living rose most rapidly. However, as the timeline in the associated boxes to this chapter shows, this recent progress was almost certainly built on earlier advances in social policy. The progress to which I am referring is that of people's standards of living in the round, from the most simple and quantifiable of measures – life expectancy and wealth – to much less measurable factors such as aspirations, hopes and fears. Much of what I am arguing is difficult to 'prove' conventionally. Instead what are presented here are conclusions derived from studying various aspects of inequality in British society – most obviously those concerning health (Shaw et al. 1999). However, this part of the chapter also argues that progress has been coupled with very little change in relative inequalities between places.

The second half of this chapter concentrates on what has not changed. I argue that, roughly speaking, the living standards of the poor lagged behind those of the rich by a generation for most of the century and continue to do so. There were few people writing at the start of the century who suggested that social inequalities as seen then would have persisted to the end. There were, of course, pessimists who saw various brave new worlds developing in which increasingly large proportions of the population would be subjugated to the will of the state. There were also utopian writers who predicted technocratic or meritocratic futures. Hardly anyone anticipated that living standards would rise for all, but rise almost equally, maintaining the overall pattern of inequalities between social groups. Even at the extremes, we now have a 'new' super-rich and a new extreme poor, mimicking (but not as great as) inequalities at the start of the century. Inequalities may have narrowed slightly, to then widen again, but in this sense we are back where we started.

The social order was not greatly reordered at any time between 1901 and 2000. Lack of progress has been expressed in the geography of British society. Those places which were poor at the start of the last century were generally still poor at its end. Almost everything has improved, but the improvement has taken place in such a way that neither the geography nor the social structure of society has been fundamentally altered. Much commentary on social change has focused on exceptions to this generalization – highlighting gentrification where it has (rarely) occurred, or searching for the growth and location of a supposed underclass. The long-term persistence of inequalities in British society has often been overlooked, as has the rapid improvement in average living standards.

British society has changed rapidly in some ways and remained remarkably rigid in others. This rigidity may not have been possible without the change, and the change may not have been possible without the rigidity. The change has been a sustained rise in the standards of living of all social groups over the last hundred years (Jones 1991). Had Britain experienced some kind of revolution at the start of the century then this little country (which relied and still relies for much of its wealth on exploiting poorer countries world-wide) may not have prospered from world capitalism as it did. Had Britain not prospered for most of the century and seen the living standards of generations rise so rapidly then the social (and spatial) structure may have been at more risk than it was. Had there not been a liberal interventionist tradition in Britain which at times helped some of the poorest places in the country to survive and which regularly improved the living standards of the poor, then Britain could have become a society as unequal as the United States.

In short I argue here that there has been great progress – and that progress allowed inequalities to continue, not to greatly widen or narrow. In making this argument I am drawing on recent collaborative work concerning trends in poverty, inequality and health, over the last two centuries, including a summary of key policy changes in boxes 2.1 to 2.4 (a truncated version of the timeline in Davey Smith et al. 2001). Here I present an argument that combines the events listed in that timeline, with the descriptions of society of contemporary observers (Miller 2000; Townsend and Gordon 1991).

Great Progress Coupled with Little Fundamental Change

Measuring social progress is only possible where there are two comparable sources of information at either end of the century. Such a situation does exist through a comparison of the 1991 census of population with Charles Booth's study of poverty in London conducted in the years leading up to 1896, published as the *Life and Labour of the People of London* in 17 volumes (Booth 1887, 1889, 1902a, 1902b). Booth's study included the *Descriptive Map of London Poverty* showing the streets of London colour-coded according to a novel social classification (Cullen 1979; Gillie 1996). Booth's classification system subsequently became the basis for the Registrar General's definition of social class, used in most subsequent population censuses. Thus, we have two points when people were allocated to social classes on a broadly similar basis and the results recorded geographically. This data was analysed to study the impacts of social material disadvantage on contemporary mortality rates (Dorling et al. 2000). Here we can look at it again in terms of social progress.

Booth and his researchers conducted a house-by-house survey of inner London in the 1880s and 1890s. They surveyed some 120,000 households, producing descriptions of household budgets, streets and classes of people (Bales 1996). This information was used to classify streets and houses by social class, and that data was then drawn meticulously onto detailed poverty maps of London (Fried and Elman 1969; Davies 1978). The *Descriptive Map of London Poverty* incorporated the sevenfold categorization of households as shown in table 2.1. Table 2.1 also gives the equivalent Registrar General's social class and the proportions of the population allocated to each social class for 1896 and 1991 within the area surveyed by Booth (see also figure 2.1).

There are striking similarities in the proportions of the population allocated to each social class despite one hundred years of social change. The three poorest groups defined by Booth match most closely with 1991 social class V. The next four classes match each census class used a century later almost exactly in terms of the character of occupations they define. Both definitions used occupation as their primary means of classifying people (or occupation of spouse or parent). The Registrar General's scheme originally included three categories of the population defined on the basis of the industry they worked in rather than their occupation, but his distinction was dropped in later classifications (Stevenson 1928: 207–30). Being a member of any particular class implied extremely different living conditions, of course, between the two time points; but the fact that such similar groups of people were grouped together in similar ways is staggering. The

Table 2.1.   Booth's classes and Registrar General's classes

Colour on 1896 map	Booth description	Equivalent Registrar General class
Black	Lowest class; vicious, semi-criminal	V
Blue	Very poor, casual, chronic want	V
Light Blue	Poor, 18s.–21s. a week for a moderate family	V
Purple	Mixed, some comfortable, others poor	IV
Pink	Fairly comfortable, good ordinary earnings	III
Red	Well to do. Middle-class	II
Yellow	Upper-middle and upper classes. Wealthy	I

Booth class	% in 1896	Census class	% in 1991
7	1.6		
6	4.0	V*	12.3
5	7.8		
4	16.6	IV	12.2
3	35.0	IIIN, IIIM	31.6
2	26.4	II	35.1
1	8.7	I	8.8

* Includes people of working age who have not worked in the last 10 years.

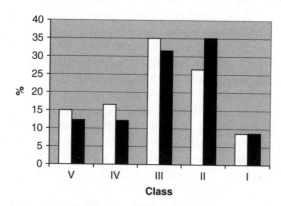

Figure 2.1.   Distribution of 1896 Booth classes and 1991 census classes
Note: white bars = 1896, black bars = 1991. See table 2.1 for definitions.

Registrar General's classification as used today is far from an anachronism, and is widely used in the advertising industry where classes I, II, IIIN, IIIM, IV and V correspond to the letters A, B, C1, C2, D and E respectively. When you hear of television advertisements being aimed at A/B audiences you are hearing an echo of the 1890s. In net terms classes D and E are slightly smaller in inner London than they were a century ago and class B slightly larger. However, the similarities between the two distributions are more interesting than the differences.

To produce the tables and figure reported above, Booth's map was digitized and placed in a Geographical Information System. Next, 1991 London ward boundaries were used as spatial units within which the distribution of the population by social class could be measured at both points in time for the same areas on the ground. These data were then further aggregated to produce the totals shown above. A relative index of poverty was then calculated to examine the changing spatial patterns of social class in London. This was based on the proportions of the population assigned to each social class in each ward (Davey Smith et al. 1998). The index ranged from 0 to 1 and was highest in areas where social class V was concentrated, and lowest in areas were there were many people in social class I. This index can be used as a measure of relative poverty at each time point, and when mapped shows how little change there was in Inner London's social structure over the twentieth century (see figure 2.2).

It is clear from the maps how little change there has been between the nineteenth-century and twentieth-century distributions of affluence and poverty in Inner London. The third map shows all-age all-cause Standardized Mortality Ratios (SMR) in these same areas in the 1990s. In the research for which these maps were originally constructed (Dorling et al. 2000) interest focused on the extent to which the 1896 map could be used to make better predictions of the spatial pattern of contemporary mortality rates than did the 1991 map. We found that all-age, all-cause SMR in the 1990s was better predicted by relative poverty as measured in 1896 than by relative poverty measured in 1991. In particular, deaths from stroke and stomach cancer (associated most strongly with childhood poverty) are better predicted by the nineteenth-century data. For other causes of death, poverty measured in 1991 provided a better predictor. The relationship with poverty in 1991 is strongest for coronary heart disease, chronic obstructive pulmonary disease, and lung cancer. At ages under 65 the 1991 census provided a better predictor for all-cause mortality. Thus the two poverty maps correlate with the pattern of mortality in a way we might expect, given what we already know about particular diseases. This suggests, among other things, that the maps are probably good general representations of social inequalities across London for their respective time periods.

For my purpose here what is most interesting about the two maps of poverty is the stability they show rather than the subtle changes. The area that has changed the most in relative terms lies south of the river around Brixton. The effect of immigration from the Caribbean was perhaps greater

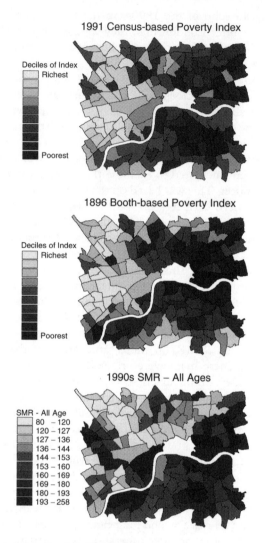

Figure 2.2.   Poverty in Inner London, 1896–1991

than the impact of a hundred years of social policy, home-building and rebuilding and two world wars (including a great deal of bombing). However, even that response was slight and its mid-century effects are now diminishing as parts of the area re-gentrify.

How representative is change in Inner London of social change more generally? There are few places in Britain where social structure was mapped over a century ago and where the results were published in a way that can be compared to the situation today. Inner London is arguably an area where one might have expected most change. Being the centre of a major world city, being most directly affected by the Blitz in the Second World War and the huge decline and then turnover (and now some growth) of population that London has experienced. However, a hundred years of social policy initiatives (many listed in the boxes below) have had almost no impact on the spatial pattern of relative poverty in inner London.

The one obvious area that can be compared to London is York, surveyed by Rowntree in 1899 (Rowntree 1901). His map was much cruder than Booth's and cannot be compared over time in the same way. Cursory comparisons of the York maps do, however, suggest that much more changed in York than in Inner London. York grew considerably over the century, unlike Inner London, and this is perhaps the most important factor in explaining the differences. The greatest differences of all in Britain are found where New Towns have been established; Milton Keynes experienced the greatest population growth in recent decades and consequently the greatest social change – but this is to compare green fields with new-build estates. What the comparison of the social geography of London shows is that where space is constrained, change too has been constrained.

Three Choices

It would be very wrong to assume that little has changed over the course of the last century in anything but the relative positions, sizes and locations of social groups in Britain. One hundred years of social change have brought about enormous progress in absolute living standards to the population of Britain. Why can I say that? Well, let me ask you a simple question. Suppose you could choose to have been born in 1851, 1901, or 1951? You are not allowed to choose the family you are born into or where you are born – just the year of your birth (such a choice following the logic most clearly set out in Rawls 1971). Imagine three upturned cups are placed before you and in each is one of those three dates. You are shown which date is in which cup. How carefully would you watch the cups being swapped around before choosing the one with your year of birth?

Had you chosen 1851, if you were lucky you would spend your short old age at the start of the twentieth century – most probably in what would be viewed now as unrecognizable squalor, cold, damp and not a little hunger and pain. Times were changing as you entered your fiftieth year, but all a little too late. Box 2.1 lists just a few of those changes. On your sixtieth birthday Parliament would pass the Contributory National Insurance Act, far too late for the spells of unemployment that you had lived through (and it wouldn't have been called unemployment then!).

Box 2.1.   Some key social policy changes in the first twenty years of the century

1901   Factory and Workshop Act dealt with health and safety, employment and education of children, dangerous and unhealthy industries, fire escapes, mealtimes, overtime, night work, and homeworking. The minimum working age was raised to 12.

1902   Midwives Act created the Central Midwives Board with responsibility for the registration of midwives, rules for their training and examination, and the regulation of their practice.

1903   Motor Car Act required numbering, registering and lighting of all cars, and introduced an offence of reckless driving.

1904   Report of the Interdepartmental Committee on Physical Deterioration made recommendations on a wide spectrum of health matters. These included: standards of purity for all food and drinks; registration of stillbirths; health visitors to be appointed in every authority.

1905   Unemployed Workmen Act permitted local authorities to set up labour bureaux to help unemployed people to find work, and to finance the voluntary emigration of those out of work.

1906   Education (Provision of Meals) Act allowed local education authorities to contribute to the provision of school meals.

1907   Notification of Births Act introduced notification of births to the medical officer of health, thus enabling the mother and baby to be visited.

1908   Old Age Pensions Act provided for a pension of 5 shillings a week for people over 70 years old with incomes of less than £31. 10.s. 0d. a year. Coal Mines Regulation Act for the first time limited the hours of work of adult males.

1909   Labour Exchanges Act led to the setting up of labour exchanges to provide information about available work. Trades Boards Act established the principle of minimum wages in defined occupations.

1910   Mines Accident (Rescue and Aid) Act provided for fire precautions, rescue work and first aid treatment at mines.

1911   National Insurance Act established health and unemployment insurance to be paid for by contributions from the state, employers and employees. This provided sickness, disablement and maternity benefits, and a medical practitioner for all insured people.

1912   Coal (Minimum Wage) Act set up district boards to fix minimum wages for workers employed underground. Trade Union Act allowed unions to use special, but not general, funds for political purposes under certain conditions.

1913 Report of the Departmental Committee on Tuberculosis recommended that dispensaries for the diagnosis and treatment of tuberculous patients and sanatorium facilities should be available.

1915 Increase of Rent and Mortgage Interest (War Restrictions) Act marked the beginning of rent control and the protection of the interests of tenants.

1916 Police, Factories Act concerned the welfare of workers in factories including the heating of workplaces; taking of meals; provision of protective clothing; seating; washing; accommodation; and the availability of first aid and ambulances.

1917 Royal Commission on the Housing of the Industrial Population of Scotland recommended that the state should accept some direct responsibility for the housing of the working classes.

1918 Representation of the People Act established a common franchise for parliamentary and local government elections and introduced enfranchisement of women aged over 30 years if they were ratepayers or wives of ratepayers.

1919 Ministry of Health Act established the ministry. In addition to its health functions the new ministry became responsible for the poor law, national insurance, local government, planning, housing, environmental health, and roads.

1920 Unemployed Insurance Act extended the scope of the 1911 act to include more than 12 million workers. Employment of Women, Young Persons and Children Act brought into effect conventions agreed by the International Labour Organization of the League of Nations, and raised the age of employment of children to 14 years.

By the time you reached 70 you would qualify for a pension – but you would be very unlikely to have lived that long. Among other things, the dangers of childbirth mean that you would have had a better chance as a man than a woman to have reached your three score years and ten. Viewed from the end of the twentieth century your life does not appear enviable. Some comfort, however, is that you would never know most of what you were missing. Only a very small proportion of the population were rich and their riches only saved them from some privations. A wealthy 50-year-old in 1901 had a far worse expectation of life than someone of that age born just 25 years later – irrespective of their social position. Put another way, the poor (widely defined) in Britain experienced the standard of living that the rich (widely defined) lived just a generation earlier. It's a gross generalization, but from health, to food, to holidays, to schooling – when comparing the best-off, say, quarter of society with the worst-off quarter – it took a generation to catch up. By which time, of course, the rich had moved on – to ever better health, food, leisure, education and so on.

Suppose you forsake the nineteenth century and opt to be born in 1901. A great war would be fought during your teenage years into which you could be unlucky enough to be drawn. In the year of your twenty-fifth birthday a general strike would be called. Life was grossly unfair, but at least people were complaining (in large numbers rather than a few agitators). You would have been obliged to bring up your family – had you had one – through the deprivations of the 1930s. But, again, at least when you were 51 you would have lived long enough to have seen a free health service established. See box 2.2 for a list of just a few of the seminal changes to

---

Box 2.2.   Some key social policy changes in the 1920s, 1930s and 1940s

1921   Marie Stopes established the first birth control clinic in north London.

1923   Matrimonial Causes Act allowed wives to petition for divorce on the grounds of their husbands' adultery.

1925   Housing Act consolidated earlier legislation relating to housing of the working classes including their sanitary conditions; maintenance of buildings; closure of houses unfit for human habitation; and improvement and reconstruction schemes. Widows, Orphans and Old Age Contributory Pensions Act introduced pensions for widows, orphans and men aged 65 and women aged 60 as from 1926.

1927   BCG vaccine against tuberculosis first used in the UK.

1928   Representation of the People Act lowered the voting age of women from 30 to 21 years.

1929   Age of Marriage Act raised the minimum age for marriage to 16 years.

1930   Housing Act and Housing (Scotland) Act made further provisions for slum clearance by local authorities, (albeit delayed by the financial crisis of 1931). Unemployment Insurance Act removed some restrictions on claiming benefit.

1934   Unemployment Act: eligibility for relief was widened but means testing on a family basis was continued. Cheap or free milk introduced on a national basis for all children at school.

1935   Housing Act defined 'overcrowding' and made it an offence, and placed an obligation on local authorities to rehouse persons from clearance areas and unfit houses scheduled for demolition. Similar provisions made for Scotland.

1936   Public Health Act consolidated the law relating to sanitation; drainage; nuisances; offensive trades; common lodging houses; water supplies; control of infectious diseases including tuberculosis; maternity and child welfare; child protection; registration of nursing homes; and the provision of hospitals. Education Act raised the school leaving age to 15 years (not enforced until 1947).

1937 Factories Act limited the hours of work of young persons under the age of 16 to 44 per week, and of those aged 16 to 18 and all women to 48; and introduced new regulations regarding lighting, heating and cleaning. Maternity Services (Scotland) Act entitled every expectant mother to have the services of a midwife and a doctor, and, if the need arose, of a consultant obstetrician.

1938 Holidays with Pay Act enabled wage regulating authorities to provide for holidays and holiday pay for workers.

1939 All infants and nursing mothers provided with fresh milk, either free or at not more than 2*d.* per pint.

1940 Food rationing (bacon, butter and cheese) introduced in January, followed by meat rationing two months later.

1941 The Minister of Health announced the government's intention to ensure the provision of a comprehensive hospital service. Introduction of purchase tax.

1942 The Beveridge Report laid the foundations of the post-war welfare state.

1943 The Medical Planning Committee of the Society of Medical Officers of Health recommended the creation of a new ministry of health.

1944 Education Act included the following: elementary schools to be replaced by infant and junior schools for primary education; secondary education to be provided free for all children in grammar, technical or secondary modern schools, selection for which was to be by an examination taken at the age of 11 (the '11-plus exam'); school-leaving age to be raised to 15 years.

1945 Family Allowances Act provided an allowance for second and subsequent children to be paid to the mother.

1946 National Health Service Act established a comprehensive health service by providing services free of charge, except where the Act expressly provided for charges; the NHS began in 1948. National Insurance Act established the welfare state on lines set out in the Beveridge Report (1942) with compulsory contributions to cover unemployment.

1947 National Health Service (Scotland) Act made provisions similar to those of the 1946 Act for England and Wales.

1948 Criminal Justice Act introduced more leniency towards criminals and virtually abolished flogging.

1949 Housing Act extended the 1936 Act to all households.

society you would have seen during the first half of your adulthood. There would still be rationing from the Second World War, but there were jobs for all by your old age. On your sixtieth birthday you saw legislation tightening up health and safety further in your workplace. A little late perhaps for you, but good for your children, and a year later an act was passed that meant your grandchildren might go to university. A year later again and plans for building more universities were made. On your seventieth birthday you had to get to grips with a new decimal currency. However, 1971 was a world away from 1921 and your predecessor's experience of old age at your age – had they reached it. Your chances of reaching 70 were very much better, but it was still not an expectation to be taken for granted. The world was changing rapidly as you aged further – perhaps not all to your taste. But you had spent your young and middle age living through some difficult, or perhaps just a little monotonous and in all probability laborious times. You had watched your country retreat from leading an empire to being in hock to the Americans and uncertainties in the flow of oil. You had had an interesting, if not easy, life.

Choose again and decide to miss out half of the last century. Now you are born in 1951. Rationing, which was generous for children, ensures you received a good diet in your formative years, generally regardless of which family you were born into. Similarly your father was almost certainly in work (and you almost certainly lived with him), you had access to free health care and better education than your parents. Your teenage years were spent in the 1960s and the government lowered the voting age to 18 on your eighteenth birthday so you could vote in the general election a year later. Box 2.3 lists some of the great changes that you saw take place in the 1950s, 1960s and 1970s. You perhaps got angry about issues like Vietnam – but it was happening a long way away from you. In your thirtieth year up to 4 million people were out of work and you could have been one of them, but more likely you were not doing too badly. The country was sharply divided. However, most people were now living on the richer rather than poorer side of the line. They, and there would be a good chance that you, showed this by voting for a Conservative government throughout your thirties and half of your forties. By the time you were 50, your home (which you were odds-on to own yourself) was worth a small fortune. Were you unlucky enough to be in the minority who missed out on this accumulation of wealth there was still a good chance that you had flown abroad for some holidays. You probably owned a car and were not too old to be daunted by technology which – from washing machines, to multi-channel television, to the internet – had transformed your everyday world. Your children received the same schooling as most other children. There were inequalities and they probably did not get to go to higher education, but you thought it inconceivable that your grandchildren would not. That

Box 2.3.   Some key social policy changes in the 1950s, 1960s and 1970s

1951 The government increased housing subsidies and pledged to build 300,000 houses per year.

1952 Housing Act raised the subsidies on house-building, encouraged council tenants to buy their houses, and allowed improvement grants to be made available to private landlords.

1954 Mines and Quarries Act consolidated legislation dealing with health, safety and welfare, and the employment of women and young persons in mines and quarries. Housing Repairs and Rent Act extended previous Acts, and set out details to be considered in defining houses as 'unfit for human habitation'.

1955 Food and Drugs Act consolidated legislation relating to the sale of food to the public, prevention of food poisoning, milk, dairies, markets, and slaughter houses.

1956 Clean Air Act introduced smokeless zones.

1957 Housing Act consolidated previous acts, setting standards of over-crowding.

1959 National Insurance Act introduced retirement pensions and contributions related to earnings.

1960 Offices Act was concerned with the health, safety and welfare of office workers. Noise Abatement Act made noise a statutory nuisance.

1961 Factories Act updated the acts of 1937, 1948 and 1959; it dealt with cleanliness; overcrowding; lighting; sanitary conveniences; safety; welfare; accidents; industrial diseases; employment of women and young persons; home work; and the duties of factory inspectors and public health inspectors in factories not using mechanical power.

1962 Education Act imposed a duty on local authorities to make grants to students who obtained places on certain courses at universities and establishments of further education.

1963 The Committee on Higher Education recommended radical changes in the structure of higher education with the doubling of student places.

1964 A free vote in the House of Commons abolished the death penalty for murder, becoming effective in 1965.

1965 Rent Act reintroduced rent control for the majority of privately owned unfurnished accommodation; gave tenants security of tenure; and introduced a scheme for the assessment of fair rents. Race Relations Act prohibited discrimination on racial grounds in places of public resort and in regard to tenancies; made incitement to racial hatred an offence; and constituted the Race Relations Board.

1967   National Health Service (Family Planning) Act enabled local health authorities to provide a family planning service for all persons. Sexual Offences Act legalized homosexual practices in private between consenting adults in England and Wales.

1968   Clean Air Act prohibited the emission of dark smoke from industrial and trade premises.

1969   Representation of the People Act lowered the age of voting to 18 years. Divorce Reform Act introduced the criterion of irretrievable breakdown of a marriage.

1970   National Insurance Act extended the eligibility for widows' pensions; introduced non-contributory pensions, as of right, to all people aged 80 years or more; and an 'Attendance Allowance' for disabled persons needing frequent or continuous attention. Family Income Supplements Act provided for a new benefit, administered by the Supplementary Benefits Commission, for families with small incomes.

1971   Education (Milk) Act restricted the duty of education authorities to provide milk for pupils. First shelter for battered wives opened.

1972   Housing Finance Act required local councils to charge 'fair rents' for subsidized council accommodation, and introduced rent rebates and allowances. Children's Act prohibited the employment of children below the age of 13 years. School-leaving age raised to 16 years.

1974   The government agreed a 'social contract' whereby the trade unions would moderate wage demands in return for promises of increased government spending on pensions, the NHS, and child benefit, price control and restricted increases in council house rents.

1975   Social Security Pensions Act introduced earnings-related retirement pensions. Sex Discrimination Act made discrimination on the grounds of sex in employment, training and related matters an offence, and established the Equal Opportunities Commission.

1976   Education Act required local education authorities to submit proposals for introducing comprehensive schooling. Commission for Racial Equality set up with a remit to promote the elimination of discrimination and the equality of opportunity for all racial groups; to support local community relations councils; and to undertake advisory and educational work.

1977   Housing (Homeless Persons) Act extended the duties of local authorities to house homeless people. Inner Urban Areas Act designated districts of deprivation for special treatment.

1980   Housing Act introduced the 'Tenants' Charter' giving council tenants the right to buy the houses they occupied.

is, if they were not among the one in six said to be living in poverty at the start of the twenty-first century. But this poverty was a world apart from that seen by your grandparents a century earlier. Things could change greatly over the coming years – but the thought that you would not live another 20 years to see these changes would have seemed overly pessimistic. Around you were people in their seventies and eighties, still enjoying life, in numbers that had never done so before.

On Progress

Let's change the game to something a little more realistic. People do not get to choose when they were born. The game of supposing that you are born into a family at random is traditionally used to ascertain people's reactions to inequalities in society at one point in time (again see Rawls 1971). Suppose that you could now choose into which family you were born in the year 2001. What kind of society would you like to be born into? One in which two-thirds were well off while a third were poor, or one which was less affluent overall but a lot more equal? Equality is the usual answer expected of players of this game. However, as it was played out in British social history for most of the twentieth century the strategy was to raise the height of safety nets for those at the bottom of society while not constraining the growing affluence of those at the top. During unusual times this was not the case. At the start of the twentieth century many of the aristocracy lost their wealth to inheritance taxation. In the middle, wartime necessity levelled the playing field significantly; and at times before the last quarter it was proposed (and occasionally implemented) that the rich should be taxed far more greatly on their incomes. But between these times wealth was permitted to beget wealth. In the last 25 years of the century the average income of the poorest quarter of society fell in relation to the best-paid quarter. Comparing box 2.4 on some of the key beneficial legislative changes in the 1980s and 1990s with the previous three boxes it is clear that government during this time was far less concerned about material inequalities and social disadvantage than before. It is extremely difficult to construct a list of government actions over the last 20 years that compares to the scale of previous achievements.

Despite the growth in inequality, by the end of the twentieth century more than 90 per cent of the poorest tenth of society had a telephone, washing machine, fridge and central heating. They almost certainly had problems paying the telephone, fuel and water bills, or stocking the fridge with good-quality food, but they actually had these things. As table 2.2 shows, the situation had changed rapidly since the 1960s. I think it is worth examining this evidence about the material goods owned by the poorest

Table 2.2.   Access to consumer durables of the bottom decile income group (%)

Percentage of individuals in household with access to a:	1962–3	1972–3	1982–3	1992–3	1999–2000
Telephone	8	20	58	78	95
Washing machine	–	54	79	89	91
Fridge or fridge-freezer	–	52	95	99	99
Car	–	26	44	56	71
Video cassette recorder	–	–	–	68	86
Central heating	–	20	46	73	90

Box 2.4.   Some key social policy changes in the 1980s and 1990s

1984   Data Protection Act regulated the use of automatically processed information relating to individuals; required the registration of data users; and established new legal rights for individuals with regard to personal data processed by computing equipment.

1985 Pasteurization of all milk for retail sale through shops and dairies became compulsory in England and Wales.

1986 Sex Discrimination Act strengthened the powers of the 1975 Act, and brought the law into line with European Community law. Protection of Children (Tobacco) Act made it illegal to sell any tobacco product to children aged under 16 years.

1988   Motor Vehicles Act made it compulsory for children to wear rear seat belts in cars.

1989   Human Organ Transplant Act prevented the commercial sale of human organs.

1995   Disability Discrimination Act made it illegal to discriminate against disabled people.

1998   Human Rights Act enshrined the European Human Rights Convention in British law; it was to come into force in 2000. National Minimum Wage Act introduced a national minimum wage.

1999   In *Opportunity for All, Tackling Poverty and Social Exclusion* the government announced plans to eradicate child poverty.

---

tenth of British people closely. By definition these people are not rich. What has happened is that all the six material goods shown in the table are now necessities to live an 'acceptable' life. A century ago the table could well have consisted of the following six items: a postage stamp, cold running water, a meat box, a bicycle, a newspaper, coal. And a century ago a commentator might have argued that some of these items were not necessary to live an acceptable life in Britain in 1901. Indeed, Rowntree in his *A Study of Town Life* of that year had to argue that being able to afford a stamp should be seen as a necessity of life (see Harris 2000):

> Let us clearly understand what 'merely physical efficiency' means. A family living upon the scale allowed for in this estimate must never spend a penny on railway fare or omnibus. They must never go into the country unless they walk. They must never purchase a halfpenny newspaper or spend a penny to buy a ticket for a popular concert. They must write no letters to absent children, for they cannot afford to pay the postage. They must never contribute to their church or chapel, or give any help to a neighbour which costs them money. They cannot save, nor can they join a sick club or Trade Union, because they cannot pay the necessary subscriptions. The children must have no pocket money for dolls, marbles, or sweets. The father must smoke no tobacco, and must drink no beer. The mother must never buy any pretty clothes for herself or for her children, the character of the family wardrobe as for the family diet, being governed by the regulation, 'nothing must be bought but that which is absolutely necessary for the maintenance of physical health, and what is bought must be of the plainest and most economical description.' Should a child fall ill, it must be attended by the parish doctor; should it die, it must be buried by the parish. Finally the wage-earner must never be absent from his work for a single day. (Rowntree 2000: 133–4)

Rowntree's phrase 'merely physical efficiency' refers to the living standards that could be attained by living just over the poverty line: a poverty line that Rowntree established in 1901. He admitted 40 years later that he had deliberately set his primary poverty line at a low level because of the adverse comments that any more generous line might have attracted: 'It was a standard of bare *subsistence* rather than *living*' (Rowntree 1941: 102). We have now made such enormous material progress in Britain that it has brought us to the point where having a video cassette recorder is increasingly viewed as a necessity (especially by parents of young children!). However, a century after Seebohm Rowntree wrote about the stamp, the foundation named after his father agonized about releasing the results of one of its surveys which did indeed suggest that a near majority of people in Britain saw having a video recorder as a mark that you belonged to society. Almost all households now feel the need for a car. Less than a third of the poorest tenth of society had no access to a car in 2000. Poor households do not run cars out of fecklessness. Massive material gains have brought about massive material reliance. Despite all this, the age-old problem remains: poverty means not having access to what is the norm. Too few people accept this. We have experienced enormous progress in some senses but almost no progress in how poverty is understood or accepted by society. The 'undeserving poor' are still viewed as such by perhaps the majority of the population and in government policy, which has recently introduced new safety nets for pensioner incomes and working families (with children), but has done far less for childless adults under retirement age – the parents and pensioners of the future.

In terms of standards of living we now live longer, eat better, are warmer in our homes and safer in our workplaces, and are better entertained. But we are also much more reliant on wealth and what it buys. We do not recognize our reliance and so we see argument after argument claiming that people who are poorer in Britain can easily live without what others take for granted. We have passed law after law after law to improve living standards and to regulate the vagaries of the free market, and constructed a welfare state that does not see people starve (except very rarely) and which now educates most people up to the age of 18 or 19 and a large number up to 21. Read again though the boxes of policy changes listed in this chapter (which you might well have been tempted to skip!) to get a rough feel for a century of change. Now, however, we turn to lack of progress – or why, despite all this apparent progress there is a feeling that little has been achieved. This is where a geographical turn is needed.

A Little More Geography

An aspatial approach to social progress can produce very positive reading. The list of progressive national achievements in boxes 2.1–2.4 makes for impressive reading, largely of Acts of Parliament, almost all of which had no direct geographical component. The systematic raising of the school-leaving age, the increased surveillance of working conditions by the state, improvements in government benefits and onwards and upwards (despite a few regressive moments not listed in the boxes). Similarly, comparing social groups can produce a quite positive, if rather misleading, picture of 'catch up', as table 2.2 suggested. And as demonstrated in table 2.1, these social groups have not greatly changed in size – although this is a contentious opinion (in particular the 2001 census results will show a huge apparent rise in the proportions of people in high classes but this will almost all be due to 'grade inflation' in job titles and a failure to redefine classification systems appropriately). However, even the widening gap between social classes in Britain in recent years appears to be narrowing. For instance, in February 2002 the Office for National Statistics reported the life-expectancy gap between social classes to be narrowing (see Hattersley 1999 for its earlier widening). A spatial approach on the other hand is often much more pessimistic and complicated. At one extreme, the current life expectancies for men living in Glasgow are roughly equal to those experienced by all men in England 40 years ago.

Table 2.3 shows the populations of five large cities in Britain and how these populations changed over the course of the last century. Glasgow contained a million people in the 1930s and 1960s. Like Liverpool, it shrank dramatically in the last third of the century. People moved between these cities and out of them at a great rate. One product of that movement, I suspect, has been to maintain the relative hierarchy of average standards of living between these places. Liverpool and Glasgow were poor places a century ago, despite being regarded as economically successful. The relative positions of the people who live in those cities, as compared to people living elsewhere, has not greatly changed. Life-expectancy tables are just part of the evidence for this. My point is that a great selective movement of people has been necessary for such relative stability in the geographical social hierarchy. The net effects of that movement on population totals are highlighted in table 2.3 (see Brimblecombe et al. 1999, 2000).

In recent years Manchester and Leeds have been said to be 'on the up'. This translates into 'have not declined as we would expect northern cities to', in population terms. The slight relative changes in city hierarchy are interesting, but too much of an obsession with small changes in rank orderings blinds us to the overall stability of the geography of society: a stability maintained through change. A century and a half ago the two English towns with the worst living conditions were probably Salford and Oldham. Their relative position has changed little since (Dorling 1997). The composition of their populations has changed dramatically to allow this, most obviously in their current ethnic make-up. Migration from around the world was necessary to maintain their social positions within England. Similarly within London, relatively poor people whose grandparents grew up in Africa, the Caribbean or the Indian subcontinent now live in areas which were disadvantaged for a very different population which preceded them (often themselves immigrants from Russia, eastern Europe and Ireland). The migration of large populations has been necessary to maintain spatial inequalities. It may well be the case that nowhere is the

Table 2.3. Population of five major cities (thousands)

	1911	1931	1961	1991	1999
London	7,256	8,216	8,183	6,890	7,285
Manchester	714	766	661	439	431
Liverpool	747	856	748	481	456
Leeds	446	483	510	717	727
Glasgow	784	1,088	1,055	663	611

influence of selective migration more important than in London. Here is a city which table 2.3 suggests is roughly the same size as it was a century ago (using the contemporary boundaries which have widened). This stability of the social class system in London is in turn reflected in a relatively stable residential hierarchy of property prices. Prices shoot up in absolute terms, but relative differentials remain quite stable. The maintenance of inequality can be seen through this geography, and this maintenance requires movement.

At a national level, the work of Gregory et al. (2001) supports the argument that social inequalities between places tended to remain much the same over the twentieth century. There were exceptions. While Newcastle and north-east England in general did unusually well in the 1960s, within London, Notting Hill did unusually badly. However, these are exceptions and in the long term what is remarkable about British social geography is its consistency. Areas that do change have tended to revert a few decades later to their former position.

We are good at spotting change, but less good at seeing stability. Compare Peter Townsend's definition of relative poverty published towards the end of the century with Rowntree's published at the start (quoted above):

> Individuals, families and groups in the population can be said to be in poverty when they lack the resources to obtain the type of diet, participate in the activities and have the living conditions and amenities which are customary, or are at least widely encouraged or approved in the societies to which they belong. Their resources are so seriously below those commanded by the average individual or family that they are, in effect, excluded from ordinary living patterns, customs and activities. (Townsend 1979: 31)

The 20 years after Townsend wrote this saw the rise of 'social exclusion' as the 'term of concern' for poverty. What Townsend describes is the same as Rowntree saw 80 years earlier and the same that contemporary commentators see now. What people are excluded from is the prosperity of others and other places – for a generation. Then, by the time they are included at those levels, general levels of prosperity have risen again: the process of 'catch up' never results in any actual catching up.

The timeline followed in boxes 2.1–2.4 goes from 1901 to 1999. This is what an extended entry for 2001 might say:

> **2001** Government figures show that income inequality increased under the Labour administration. Tony Blair is re-elected as prime minister (Labour), but with a record low turnout of the electorate and widespread apathy. On 11 June Mr Blair awarded himself a 40 per cent pay increase, raising his salary to £163,000 a year. In an interview a week earlier in response to the question 'Is

it acceptable for the gap between rich and poor to get bigger?' he would only answer by saying 'It is acceptable for those people on lower incomes to have their incomes raised'. He was not concerned about the gap between 'the person who earns the most in the country and the person that earns the least'. Research conducted on behalf of the Joseph Rowntree Foundation finds that 2 million children in Britain – more than one in six – are experiencing multiple deprivation and poverty.

Conclusion

The central argument of this chapter is: that the twentieth century in Britain saw enormous social progress in absolute terms, but remarkable social rigidity in the basic structures of society, particularly in its inequities. This rigidity is reflected through a geography in which the relative positions of people living in different cities or different parts of cities tended to remain the same. The maintenance of inequality requires a great deal of change. This change has many components. One component is the large-scale systematic migration of people to maintain spatial hierarchies. For instance, those people who can, do leave poorer areas, while those who are well-off clamour to spend even more of their salaries to live in still richer places. There needs to be almost continuous absolute improvement in average wealth for continued inequalities between social groups to be accepted by the majority. Inequalities can be maintained as long as everyone is getting a little richer. Inequities in British society have been maintained while most have become richer. At the start of the twenty-first century there is little to suggest this process will not continue for some time.

ACKNOWLEDGEMENTS

Danny Dorling is very grateful to his undergraduate tutees of 2002/3, who helped to search for key positive policy changes which have occurred during their lifetimes to insert in box 2.4. Unfortunately they, like him, were unable to expand the list greatly.

REFERENCES

Bales, K. 1996: Lives and labours in the emergence of organised research, 1886–1907. *Journal of Historical Sociology*, 9(2), 113–38.
Bartley, M., D. Blane, and J. Charlton 1997: Socioeconomic and demographic trends, 1841–1994. In J. Charlton and M. Murphy (eds), *The Health of Adult Britain 1841–1994*. London: The Stationery Office.

Booth, C. 1887: The inhabitants of Tower Hamlets (School Board Division) their condition and occupations. *Journal of the Royal Statistical Society*, 50, 326–401.

Booth, C. 1889 [1969]: *Life and Labour of the People. First Series, Poverty (I) East, Central and South London*. London: Macmillan.

Booth, C. 1902a [1969]: *Life and Labour of the People. First Series, Poverty (II) Streets and Population Classified*. London: Macmillan.

Booth, C. 1902b: *Life and Labour of the People in London. Final Volume. Notes on Social Influences and Conclusions*. London: Macmillan.

Brimblecombe, N., D. Dorling and M. Shaw 1999: Mortality and migration in Britain – first results from the British household panel survey. *Social Science and Medicine*, 40(7), 981–8.

Brimblecombe, N., D. Dorling, and M. Shaw 2000: Migration and geographical inequalities in health in Britain: an exploration of the lifetime socio-economic characteristics of migrants. *Social Science and Medicine*, 50(6), 861–78.

Cullen, M. 1979: Charles Booth's Poverty Survey: some new approaches. In T. C. Smout (ed.), *The Search for Wealth and Stability: Essays in Economic and Social History presented to M. V. Flinn*. London: Macmillan.

Davey Smith, G., D. Dorling, and M. Shaw,(eds) 2001: *Poverty, Inequality and Health: 1800–2000 – A Reader*. Bristol: The Policy Press.

Davey Smith, G., C. Hart, D. Blane and D. Hole 1998: Adverse socio-economic conditions in childhood and cause-specific adult mortality: prospective observational study. *British Medical Journal*, 316, 1631–5.

Davies, W. J. D. 1978: Charles Booth and the measurement of urban social character. *Area*, 10(4), 290–6.

Dorling, D. 1997: *Death in Britain: How Local Mortality Rates have Changed: 1950s–1990s: Technical Report*. York: Joseph Rowntree Foundation.

Dorling, D., R. Mitchell, M. Shaw, S. Orford and G. Davey Smith 2000: The Ghost of Christmas Past: the health effects of poverty in London in 1896 and 1991. *British Medical Journal*, 321, 1547–51.

Fried, A., and R. M. Elman 1969: *Charles Booth's London: A Portrait of the Poor at the Turn of the Century, Drawn from His 'Life and Labour of the People in London'*. London: Hutchinson.

Gillie, A. 1996: The origin of the poverty line. *Economic History Review*, 49(4), 715–30.

Goodman, A., P. Johnson and S. Webb (eds) 1997: *Inequality in the UK*. Oxford: Oxford University Press.

Gregory, I., D. Dorling and H. Southall 2001: A century of inequality in England and Wales using standardized geographical units. *Area*, 33(3), 297–311.

Harris, B. 2000: Seebohm Rowntree and the Measurement of Poverty, 1899–1951. In J. Bradshaw and R. Sainsbury (eds), *Getting the Measure of Poverty: The Early Legacy of Seebohm Rowntree*. Aldershot: Ashgate, 60–84.

Hattersley, L. 1999: Trends in life expectancy by social class – an update. *Health Statistics Quarterly*, 2, 6–24.

Jones, K. 1991: *The Making of Social Policy in Britain 1830– 1990*. London: Athlone Press.

Miller, G. 2000: *On Fairness and Efficiency: The Privatisation of the Public Income over the Past Millennium*. Bristol: The Policy Press.

Mitchell, B. R. 1988: *British Historical Statistics*. Cambridge: Cambridge University Press.

Office for National Statistics 2000: *UK 2000 in Figures*. London: The Government Statistical Service.

Rawls, J. 1971: *A Theory of Justice*. Cambridge, Mass.: Belknap Press of Harvard University Press.

Rowntree, B. S. 1901 [repub. 2000]: *Poverty: A Study of Town Life*. Bristol: Policy Press

Rowntree, B. S. 1941: *Poverty and Progress*. London: Longman Green.

Stevenson, T. H. C. 1928: The vital statistics of wealth and poverty. Part II. *Journal Royal Statistical Society*, 91, 207–30.

Shaw, M., D. Dorling and N. Brimblecombe 1999: Life chances in Britain by housing wealth and for the homeless and vulnerably housed. *Environment and Planning A*, 31, 2239–48.

Townsend, P. 1979: *Poverty in the United Kingdom*. Harmondsworth: Penguin.

Townsend, P., and D. Gordon 1991: What is enough? New evidence on poverty allowing the definition of a minimum benefit. In M. Adler, C. Bell, J. Clasen and A. Sinfield (eds), *The Sociology of Social Security*. Edinburgh: Edinburgh University Press, 35–9.

Chapter Three

The Conservative Century? Geography and Conservative Electoral Success during the Twentieth Century

Ron Johnston, Charles Pattie, Danny Dorling and David Rossiter

> The Conservative party dominated British politics to such an extent during the twentieth century that it is likely to become known as 'the Conservative century'. Either standing alone or as the most powerful element in a coalition, the party held power for 70 of the 100 years since 1895. For much of the remaining 30 their opponents had only a fragile grip on office. (Seldon and Ball 1995: 1)

Political parties are formed to fight and win elections, and on this criterion the Conservative party was by far the most successful in Great Britain during the twentieth century. Although two of its worst performances came at the century's beginning and end – it gained 43.6 per cent of the votes cast but only 23 per cent of the seats in 1906,[1] and 30.7 and 25.7 per cent respectively in 1997 – between those dates its election machine delivered victories which were unmatched by its opponents,[2] in part no doubt because throughout the century it was best able to raise the money necessary to fight national and local campaigns (Pinto-Duschinsky 1981).

The Conservative party won the largest percentage of the votes cast at 17 of the 26 elections, won the largest percentage of the seats at 14 (which in one case was less than a majority), and was in office – either alone or in coalition – for 66 out of the 100 years (figure 3.1). This success rate was a significant improvement on its nineteenth-century performance: between 1832 and 1885 it won only two of the 13 elections outright, compared to eight of the 14 between 1886 and 1935 and eight of the 16 between 1945 and 2001: in a period of rapid social change, it has dominated British politics (Pattie and Johnston 1996) – although it won the support of more than half the electorate on only three occasions. Why this success? Why did the Conservative party so dominate British government for most of that century?

Exaggeration and Bias in the Electoral System

Explorations of the Conservatives' success have to take account of the geographies involved, as well as the contexts within which each general election was held. In the British electoral system, using single-member constituencies (with a small number of exceptions up to 1949), the candidate having the largest number of votes is elected, even if he/she obtains less than 50 per cent of those cast. Disproportionality in election results in the translation of votes into seats is the norm with such systems. This disproportionality has two components: *exaggeration*, whereby one or more parties (usually the largest) gets a larger share of the seats than of the votes cast (a 'winner's surplus'); and *bias*, whereby one of the two largest parties gets a larger proportion of the seats than the other, with the same percentage of the votes.

Exaggeration was normal throughout the century, with each of the two largest parties getting a bigger percentage of the seats than of the votes cast at most of the contests. This can be shown by the seats:votes ratio (a party's percentage of the seats allocated divided by its percentage of the votes cast: a ratio above 1.0 indicates a larger percentage of seats than votes). For the Conservatives, this occurred at 20 of the 26 elections analysed (figure 3.2), all of those at which it was the largest party and an additional five when it came second in the contest for votes. Its largest seats:votes ratios occurred in three periods: in the 1920s and 1930s, when the opposition was divided between two parties, Labour and Liberal; in the 1930s after MacDonald's formation of a National Government in 1931 resulted in the Conservatives' best two performances in terms of seats won; and in the 1980s, when again the opposition was split between two parties, neither of which was particularly strong.

The disproportionality associated with first-past-the-post electoral systems led to the Conservatives almost invariably winning a larger share of the seats than of the votes, but this was the case for its main opponents also at most of the elections. But did the system's disproportionality favour the Conservatives over the Liberals until the 1920s and over Labour thereafter? Answering this question involves looking at the degree of bias in a result: that is the difference in the number of seats obtained by the two parties if they had the same percentage of the votes cast. This is calculated by applying a uniform shift in votes across all constituencies to equalize the

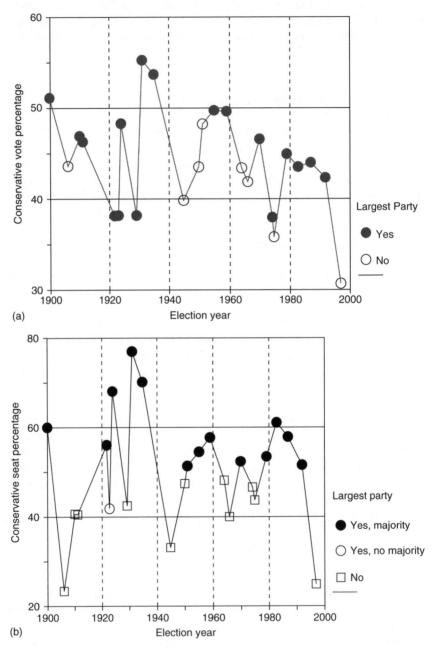

Figure 3.1. Conservative percentages of (a) votes cast and (b) seats won at twentieth-century general elections (excluding 1918)

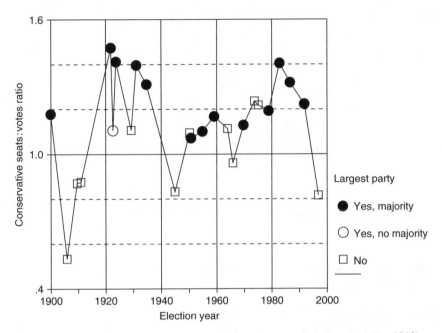

Figure 3.2. Conservative seats:votes ratio at twentieth-century general elections (excluding 1918)

parties' vote totals. Thus, for example, if the Labour party won 45 per cent of the votes nationally and the Conservative party won 35 per cent, the Labour share would be reduced by 5 percentage points across all constituencies, and those votes re-allocated to the Conservatives, so that each had 40 per cent of the total (Johnston et al. 2001). With the same vote percentages, an unbiased system should have each party with the same number of seats. Any difference indicates the level of bias. Figure 3.3 shows that the bias in the system (i.e. a positive value for the bias figure) favoured the Conservatives at 14 of the elections, including 10 of those contested after 1940.

The British method of translating votes into seats that generates both exaggeration and bias occurs because of the interaction of two geographies. The first is the geography of voting, of who votes what, where. This involves maps which, at their most detailed, show the home location of each elector and how he/she voted.[3] More generally, however, these point maps are generalized into area maps, showing the pattern of support for each party (and of abstentions) by wards and other administrative divisions. Secondly, there is the geography of the constituencies, the areas which return the MPs to Parliament. This geography is overlaid on the geography of voting, with local government wards used as the building-blocks by Boundary Commissions for creating constituencies.

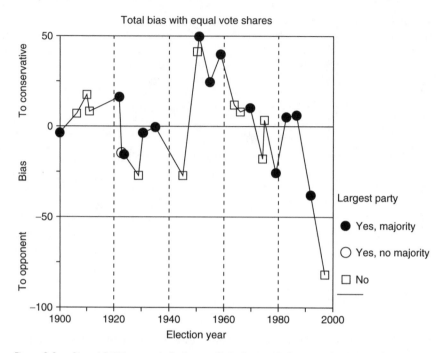

Figure 3.3.   Bias at British general election results in the twentieth century (excluding 1918)

The particular set of constituencies interacts with the geography of voting to produce areas of strength and weakness for each party – as well as the arenas within which the parties seek to mobilize support. How the constituencies are created is thus a crucial influence on the translation of votes into seats, affecting both the level of disproportionality in any election outcome and, potentially, any bias between parties in how they are affected. There are many different ways of producing a set of constituencies from a larger number of wards: Johnston and Rossiter (1982), for example, identified over 15,000 ways in which the 27 wards in Sheffield could have been amalgamated into six constituencies during the redistribution completed in 1983.

During their successful century, therefore, the Conservatives benefited from the distortions produced by the electoral system at a majority of the elections fought – and at a preponderance of those fought after the Second World War: the biases within the system favoured them rather than their opponents. Those biases are a consequence of the geographic framework within which elections are conducted. Understanding how they operated in the Conservatives' favour for much of the twentieth century is the focus of the rest of this chapter.

The Changing Electoral Context

The impact of these geographies changed considerably across the twentieth century, as the electoral context was altered. Three aspects of this context were particularly important. The first was the extension of the franchise, a process initiated with the Great Reform Act of 1832 and continued by the further Reform Acts of 1866 and 1885. Even after the latter, however, the majority of adults (including all adult females) were unable to vote at general elections; in 1910, for example, of an adult population of 26.1 million only 29.5 per cent were enfranchised. After the 1918 Representation of the People Act, when all adult males together with females aged over 30 could vote, the percentage increased to 79.9, and the Equal Franchise Act 1928 generated a further increase to 91.1 per cent. By 1939, 98.5 per cent of the country's adult population was on the electoral roll. Other aspects of the franchise have influenced turnout and voting patterns notably, though not only, at the elections held in wartime or just after hostilities ceased (only one of which – 1945 – is analysed here). In the mid-1980s, for example, the government legislated to allow those living abroad to vote at UK elections for 20 years after their departure (in the constituency where they last resided): against this, the legislation introducing the 'community charge' for funding local government (commonly known as the 'poll tax') led a substantial number to disqualify themselves by failing to register as electors in the hope that this would help them to evade the charge (on these, see Dorling et al. 1996; Pattie et al. 1996).[4]

These changes in the size of the enfranchised population had major implications for the political parties, because each extension up to 1918 increased the proportion of the electorate who were in what was generally termed the 'working class', much of whose support was mobilized by the Liberal and, increasingly, Labour parties. It should be noted, however, that the Conservatives were successful in winning working-class support in some places (McKenzie and Silver 1968; Nordlinger 1967). The class composition of the British electorate changed and, given that class was a dominating feature of electoral mobilization throughout most of the century, this should have disadvantaged the Conservative party. Elsewhere in western Europe, extension of the franchise led many conservative parties to support the introduction of proportional representation systems in order to sustain their electoral position: this was not the case with the British Conservatives, however, who remained committed to the 'first-past-the-post' system and, as we have seen, thrived electorally.

The franchise extension was in itself geographically constituted, in that the working-class residents who gained the right to vote in the late nineteenth and early twentieth centuries were predominantly urban. This had clear implications for the parties' electoral prospects, if they could mobilize the new voters in such places. This was of major importance to the Conservatives, since increasingly the urban areas became dominated by voters whose likely political affiliations would favour parties other than theirs; the class cleavage increasingly became an urban–rural cleavage, and as the urban areas grew in relative importance, so the Conservatives' 'natural' electoral heartlands apparently declined.

Whereas the first half of the century saw the enfranchisement and growth of the working-class electorate, the second half, and especially the last quarter, saw its decline as many traditional manufacturing industries were run down, a process that was countered by a growth in service industries – most of them in different parts of the country from those where manufacturing had traditionally flourished. This change in the country's class composition favoured the Conservatives, especially in the areas where service-sector growth was strongest. As a consequence, Crewe (1986) argued that for Labour to regain power in the late twentieth century it had to abandon its almost exclusive focus on economic and social policies aimed at working-class constituencies and instead target a much wider and more diverse audience embracing substantial sections of the middle class. Neil Kinnock started such a project in the 1980s; it was carried forward by John Smith (1992–4) and then successfully completed by Tony Blair after 1994. It had seemed for a while in the 1980s that a new middle-of-the-road alliance of Liberals and Social Democrats would break the two-party mould of British politics established in the 1920s when Labour replaced the Liberals as the radical party of the left (Crewe and King 1995). That project failed, however, and the late 1990s saw a very substantial Labour revival linked to policy changes which took it into the centre ground of British politics, previously occupied by the Liberal–SDP Alliance, and to the right of the current Liberal Democrat party (Budge 1999).

The second, very much related, component of the changing electoral context was the massive redistribution of population through the country. The urban component increased significantly through the nineteenth century, largely because of net migration from the countryside. The traditionally Conservative rural areas were depopulated, with many of the migrants joining the newly enfranchised urban working class. By the third decade of the twentieth century, the urban areas dominated. After the Second World War, however, this trend was partly reversed, although the country remained dominantly urban – to what extent depending on how urban was defined. Throughout the century, capitalizing on developments in both public and private transport, the middle classes moved out from the inner cities to lower-density suburbs, where they were joined to some extent by working-class residents who were provided with out-of-town council homes. From the 1960s on, the centrifugal forces were increasingly propelling people beyond the suburbs into the smaller towns of the surrounding rural areas (Champion 1998). Through counter-urbanization, the traditionally pro-Conservative areas were being repopu-lated, largely by the middle classes who were most likely to support that party, whereas the anti-Conservative urban areas were suffering the sort of population denudation that the rural areas experienced earlier in the century.

The third aspect of the changing political context – already alluded to – was the changing organization to the left of the Conservatives on the political spectrum. For the first quarter of the century, their main oppos-ition was provided by the Liberals, who won three elections between 1906 and 1918 – although only just in 1910, after challenging the Conservative hegemony in the House of Lords and as difficulties arose over their pro-posals for Home Rule for Ireland, which won it parliamentary support from various Irish parties. But the slow advance of support for Labour, especially in the industrial areas where it mobilized most actively, and the split in the Liberal party following its coalition with the Conservatives during the latter part of the First World War, saw a major switch in partisan support on the left. By the 1929 election, the Liberals were down to 23 per cent of the votes cast – which delivered them just 59 seats – whereas Labour won 37 per cent of the votes and 288 seats. This new bipolar axis to British party politics was submerged by the events of the early 1930s, MacDonald's response to the economic crisis involving a coalition with the Conservatives and a split in his own party. But it re-emerged in 1945, and – apart from a brief period in the early 1980s when the Alliance of Liberals and Social Democrats briefly led in the opinion polls and threatened to push Labour into third place at the 1983 general election (in terms of votes if not seats) – it remained firmly in place for the second half of the century.

In terms of the country's class composition and the franchise, therefore, the century's early decades were pro-Conservative, as was also the case in the 1980s. But in the middle decades radical parties of the left were more favoured by the country's demographics. In the 1930s they failed to realize that potential, following the debacle of the MacDonald Labour government elected in 1929. From 1945 to 1979, however, they won six of the 11 elections – although after the party lost three in a row in 1951, 1955 and 1959 some wondered whether Labour would ever win again (Abrams, Rose and Hinden 1960). So why were the Conservatives the dominant force in UK politics, and by far the most successful electorally, for so much of the century? In part this reflected the electoral context: the replacement of the Liberals by Labour in the 1920s, the creation of the National Government by MacDonald in 1931, and the 'You've never had it so good' prosperity delivered by Macmillan's government in the late 1950s, when the Labour party was badly split, as it was again in the 1980s. Were the Conservatives aided further by the operation of the electoral system, thereby creating the large Conservative seats:votes ratios in the 1920s and 1930s (figure 3.2) and the pro-Conservative bias between 1935 and 1964 (figure 3.3)? And was that assistance serendipitous or planned for?

Creating the Interacting Geographies: Constituency Delimitation

Some of the various trends outlined above favoured the Conservative party's campaigns for electoral support, whereas others operated somewhat to its disadvantage. But the extent of their impact depended on one of the key geographies introduced above – that of the constituencies. Different sets of constituencies in an area can lead to significantly different election results – even if the geography of party support in the building-blocks remains unchanged. The choice of when and how to change constituencies can be crucial to a party's chances in the translation of votes into seats.

The process of defining new constituencies – what the British call redistribution and the Americans redistricting – is open to abuse, as it is possible to so arrange the geography that one party is significantly advantaged. Two main abuses are widely recognized, notably, though not only, in the United States. The first is *malapportionment*, whereby constituencies of different size are created so that one party is strongest in areas with small constituencies (i.e. have a low average number of voters per constituency) whereas its opponent(s) are strongest in the areas with larger constituencies (on international patterns of malapportionment, see Samuels and Snyder 2001). A consequence of this difference is that it takes fewer votes to win a seat in the areas with smaller than with larger constituencies, so that the strong party in the former gets a greater return for its votes.

Malapportionment may be a deliberate strategy employed by those involved in constituency delimitation, or it may be that differences in constituency size determined for other reasons serendipitously favour one party over another. Furthermore, the distribution of voters across a set of constituencies is likely to change over time, as with the urbanization of the first half of twentieth-century Britain followed by counter-urbanization thereafter, and the patterns of population growth and decline may favour one party over another, in what might be termed *creeping malapportionment*. A party with major strengths in the rural areas would be advantaged by urbanization, therefore, as population decline there reduced the average constituency size, while increasing it in the urban areas that were attracting the immigrants. Creeping malapportionment can be halted by a redistribution, by redrawing the constituency map so as to remove the electoral inequalities induced by population movements.

The second major form of electoral abuse is *gerrymandering*, whereby – even if constituencies are equal in size – their boundaries are carefully delineated so as to create as many constituencies as possible that will be won by the party involved, or by the party supported by those in charge of the redistribution. Again, the strategic goal is to create an advantage for one party, giving it a better return on its votes won relative to its opponents. Gerrymanders come in two main forms. The stacked gerrymander involves the concentration of one party's voters into some constituencies, which it wins by very large majorities – with many of its votes being surplus (i.e. do not bring returns in terms of seats won). The cracked gerrymander, on the other hand, involves so distributing a party's votes that it wins as many constituencies as possible, in most cases by fairly small margins. It therefore wastes relatively few votes on seats that it loses, but opens itself to defeat in many of those that it wins if there is a swing of support away from it.

There is a major difference between these two strategies – only one can be countered by explicit criteria. Malapportionment can be made illegal by requirements that all constituencies be the same size (or as equal as is practicable) – as the United States Supreme Court required in a series of landmark decisions in the 1960s – and creeping malapportionment can be reduced in its impact, though not totally removed, by regular redistributions, either at defined intervals or when the differences between constituencies exceed a pre-set threshold. Gerrymandering, on the other hand, cannot be eliminated from electoral systems which use the first-past-the-post method in single-member constituencies. Explicit gerrymandering by interested parties may be prohibited by preventing them from taking part in the redistribution process – as with the use of independent Boundary Commissions in the United Kingdom since 1945 – but, as Taylor and Gudgin (1975: 405–15) put it, 'all districting is gerrymandering'. The process of combining building-blocks into larger territorial units is, as they showed, almost certain to advantage one party over another and thereby produce the equivalent of a gerrymander – an unintended gerrymander (Gudgin and Taylor 1979). In less pejorative terms (because gerrymandering implies corruption to some people, even though it was ruled entirely constitutional by the United States Supreme Court in 2001) the party benefiting from the gerrymander does so because its votes are more efficiently distributed than its opponents': fewer of them are either piled up as surplus votes in its safe seats or wasted in seats that it loses by relatively small margins.

Redistribution in the United Kingdom

So, has there been malapportionment and gerrymandering in the United Kingdom and, if so, have they contributed to the Conservative party's electoral success over the twentieth century? To answer that, it is necessary to appreciate the redistribution process, which was changed very significantly at mid-century.

Although the British electorate increased sevenfold during the nineteenth century, and population redistribution from rural to urban areas was extensive, there were only three redistributions of parliamentary constituencies then, each related to the major Reform Acts which extended the franchise. From the outset, there were massive variations in constituency size, with many boroughs (mainly small rural towns and villages – the 'rotten boroughs') having electorates of less than 50. A major goal of each redistribution was to eliminate the smallest constituencies and achieve some level of equality. All of the major decisions were taken by Parliament, however, and so party considerations were uppermost in many minds. Furthermore, as seats were reallocated from the rural areas to the burgeoning industrial cities, so boundaries had to be drawn to encompass the new constituencies – and in this there was a great deal of gerrymandering, notably by Lord Salisbury, leader of the Conservative party at the time of the 1885 redistribution, who had also been much involved in the preceding redistribution of 1866. Through detailed knowledge of the geography of voting, he was able to bargain with the Liberals for constituency boundaries which protected Conservative interests at a time when both franchise extension and population movements were running in a counter-direction (Salisbury 1884; Roberts 1999; Rossiter, Johnston and Pattie 1999).

The first twentieth-century redistribution did not take place until 1919. In the interim, the Conservatives under Balfour had prepared for a redistribution in 1905, the main outcome of which would have been a reduction of seats in Ireland, from where the Home Rule parties traditionally supported the Liberals in the House of Commons, and an increase of about 20 for England, many of which could well have been won by the Conservatives – especially as there was some gerrymandering of boundaries 'in order to secure a fairer grouping of areas in accordance with the general character of the population' (Rossiter, Johnston and Pattie 1999: 46–9). The government fell in 1906, however, and lost the subsequent election by a massive margin. The Conservatives had a large majority in the House of Commons after the 1900 election, but the party was badly split over the issue of tariff reform (figure 3.1b).

In the redistribution of 1917–19, a complicated matter involving several Acts, produced by a Liberal-led coalition, the detailed task of defining boundaries was delegated to commissions. Their prescribed goals embraced equality in electorates across constituencies, but there was also implicit gerrymandering, with the instructions including the requirements that: 'regard shall be had to ... size, to a proper representation of the urban and rural population, and to the distribution and pursuits of such population' (Rossiter, Johnston and Pattie 1999: 50).

Most of the new constituencies had populations of between 50,000 and 90,000, but variation within those limits was considerable. Within little more than a decade, creeping malapportionment was very substantial, and there was parliamentary pressure for a further redistribution in the mid-1930s, which the (Conservative) government declined to undertake.

The next, and by far the most significant, change came in 1944 as part of preparations for the post-war elections. A committee chaired by the Registrar General was established in 1942 to explore the many issues involved in modernizing the electoral system, with terms of reference including the redistribution process. Its report set out a series of principled recommendations, which included the need for regular reviews, for equality of electorates across constituencies, for use of the local government map as the template within which the geography of constituencies should be set, and for the appointment of independent commissions to undertake the reviews (Rossiter, Johnston and Pattie 1999). These were very largely accepted by a Speaker's Conference which reported in 1944, and were implemented in the, largely uncontroversial, House of Commons (Redistribution of Seats) Act 1944. This was amended a number of times in successive decades. The anti-malapportionment clause was weakened early on, by a Labour administration. Inter-country malapportionment was introduced by the Conservatives in 1958 with guaranteed minimum numbers of seats for Scotland and Wales irrespective of population changes, alongside an amendment which reduced the need for reacting to 'mild' creeping malapportionment, in order to retain constituencies in which 'community ties' had been built. Some potential for gerrymandering was also incorporated in 1958 with the opportunity for political parties (and others) to contest the commissions' recommendations at public inquiries. However, the main features of the system remained in place for the remainder of the century, being consolidated in the Parliamentary Constituencies Act 1986.

Identifying Electoral Bias

The Conservative century was therefore a game of two halves with regard to the procedures for defining parliamentary constituencies. How was the party affected by them, and how did it operate within them? To answer these questions, we apply methods that have previously been used to chart the nature of bias in the United Kingdom electoral system since the new procedures were put in place in 1944 (Johnston et al. 2001).

Effective, wasted and surplus votes

The goal for those involved in deliberately manipulating a constituency map for partisan gain, as described above, is to maximize the impact of their votes – i.e. to get as many seats as possible for a given percentage of the votes – and to minimize the impact of their opponents' votes. We can illustrate this by dividing all votes cast for a party into three types:

1   *Effective votes* – those that help the party to win seats.
2   *Surplus votes* – those that are surplus to requirements in the seats that it wins, and so do not help to deliver seats.
3   *Wasted votes* – those won in constituencies where the party loses, and so play no part in delivering seats.

Thus in a constituency with 10,000 voters (none of whom abstain), where party A gets 6,800 votes and party B gets the remaining 3,200, all of B's votes are wasted. For A, 3,201 votes are needed for victory over B (i.e. are effective), and the remainder (3,599) are surplus. If the electoral system and the geography of votes are even-handed to both parties, then each should have a similar proportion of effective, wasted and surplus votes.

To test whether this was the case through the twentieth century, we have calculated the number of effective and surplus votes per seat won at each election for the Conservative party and its main opponent (Liberal before 1924, Labour thereafter) and the number of wasted votes per seat lost – on the assumption that each party won the same proportion of the national vote total.[5] We then expressed each figure for the Conservatives as a ratio of that for its main opponent. If the ratio of the two figures were 1.0, then there would be no malapportionment or gerrymandering-like bias effects: if it were greater or less than 1.0, one party would be advantaged over the other by the interaction of the two geographies. Figure 3.4a–c shows the trends for each of these ratios across the twentieth century.

Parties are disadvantaged if they win too many surplus votes – i.e. they have larger majorities in the seats that they win than their opponents do where they win, their seats:votes ratios are relatively small. In figure 3.4a a ratio of less than 1.0 indicates that the Conservative party is advantaged on this component (i.e. it had fewer surplus votes per seat won than Labour), which indeed was a characteristic for much of the century. Only in the period 1920–45 was it common for the Conservatives to win with larger majorities on average than their opponents – no doubt a reflection of the

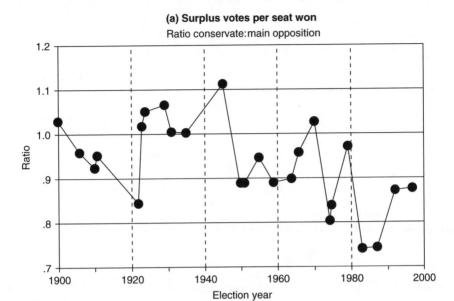

(a) Surplus votes per seat won

Ratio conservate:main opposition

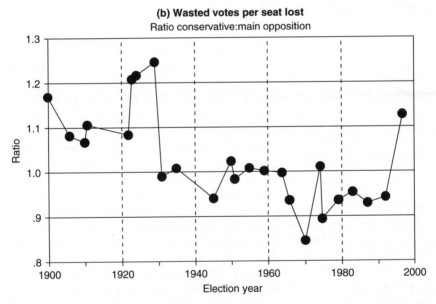

(b) Wasted votes per seat lost

Ratio conservative:main opposition

Figure 3.4. Ratio between Conservative and main opposition party's (a) surplus votes per seat won, (b) wasted votes per seat lost and (c) effective votes per seat won at twentieth-century general elections (excluding 1918)

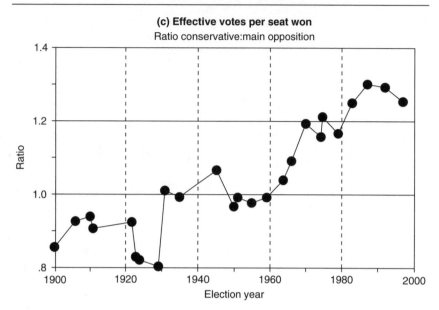

Figure 3.4. *Continued*

inter-party contest on the left in the 1920s and, especially, the Labour party's problems in the early 1930s. With wasted votes, a ratio greater than 1.0 indicates that the Conservatives tend to waste more votes where they lost than their opponents did where they lost. Figure 3.4b shows that this was common after the mid-1920s: up until then, the Conservatives tended to waste more votes than their opponents; from then until the 1960s there was little difference between the two, but (with the exception of the first 1974 election) the next three decades were characterized by ratios below 1.0, indicating that Labour wasted on average more votes in the seats that it lost than the Conservatives did where they lost. The 1997 election saw a further shift, however, with the advantage switching to Labour – an advantage that was extended in 2001 (Johnston et al. 2002).

The trend in effective votes (figure 3.4c) shows that the century's last four decades involved a substantial, and growing, anti-Conservative bias. Until the mid-1920s, the Conservatives had many fewer effective votes per seat won than the Liberals, suggesting that the Conservatives were strongest in the country's smaller constituencies, where fewer votes were needed for victory. In other words, because of the geography of their support, it took fewer votes to elect a Conservative MP than one of their opponents: these Conservative votes, of course, were concentrated in the depopulating rural areas. From the 1930s until the mid-1960s there was little difference between the two parties, but from then on the Conservative disadvantage was compounded: by the early 1990s they needed on average nearly 30 per cent more votes to win a seat than did Labour (i.e. ratios of *c*.1.3). The rural areas were being repopulated and the inner cities depopulated, to the Conservatives' detriment in electoral battles.

This interpretation is clarified by figure 3.5, which shows the trend in the ratio of average electorate in seats won by Conservatives to those won by its opponents: a ratio below 1.0 indicates that Conservative-held constituencies were smaller than Liberal-held seats (before 1920) or Labour-held seats (after 1920). At the century's first four elections, the Conservative strength in the declining rural heartlands gave it a significant advantage, which it had lost by the mid-1920s and which, from then until 1945 when many urban constituencies were small because of the consequences of the war, decentralization removed. For a short period in the 1950s the Conservatives regained the advantage, because the Boundary Commission for England operated – under a Conservative government – a policy of favouring the rural areas with slightly smaller constituencies than the urban, which it ended in the 1960s. The remainder of the century had the advantage very much with Labour, with the peak differences in 1970, 1979 and 1992 reflecting the impact of creeping

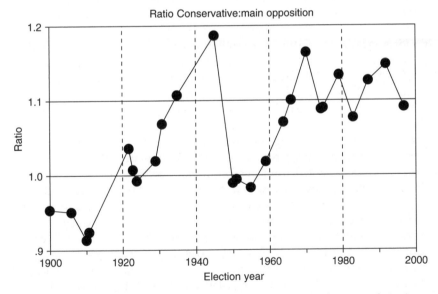

Figure 3.5. The ratio between Conservative and the main opposition party's average electorate per seat won at twentieth-century general elections (excluding 1918)

malapportionment. Commission reviews after each of those elections reduced the size variation between constituencies, largely by reducing the number in (the mainly pro-Labour) urban areas and increasing the number in the rural districts.

Bias and its components

Malapportionment and creeping malapportionment apparently characterized the British electoral system both before and after the reforms of 1944, therefore; but how important were they to Conservative success, both in absolute terms and relative to gerrymandering-like effects? To answer that, we assess bias with the simple metric introduced earlier – the difference in the number of seats that the two parties would win if they had the same share of the votes cast – which has the added advantage of being decomposable into malapportionment, gerrymandering (or efficiency) and other components. In these calculations, a positive bias indicates that the Conservatives would have won more seats than their main opponent with equal vote shares, with the reverse situation shown by a negative figure.

The overall trend in the volume and direction of bias, as already indicated (figure 3.3), shows that the Conservatives were the main beneficiaries of the electoral system's vagaries in translating votes into seats in two main periods: the first two decades, and the period from 1950 to 1970. In the intervening period, the system worked somewhat to the Conservatives' disadvantage, although the difference was never more than 40 seats. And then from the late 1960s onwards, the party's advantage (of up to 50 seats at its peak) was rapidly eroded, until 1997 when it suffered a disadvantage of 82 seats. In 2001 the figure was even greater at 141.

What part of this pattern relates to malapportionment? Figure 3.6a shows the volume of bias due to the malapportionment component, which measures the impact on the number of seats won (at equal vote shares) of variations in electorate size. In the first quarter of the century, the Conservatives benefited from this component because of their strength in the depopulating rural areas, but as the inner cities began to decline in population so the bias shifted to favour Labour. Since the implementation of the House of Commons (Redistribution of Seats) Act 1944, however, one of the main purposes of the Boundary Commissions' regular reviews has been to counter growing variations. Creeping malapportionment restarts after each review, with new constituencies used at the general elections of 1955, 1974 (February), 1983 and 1997. Each review reduced the pro-Labour malapportionment bias, which increased again at subsequent elections.

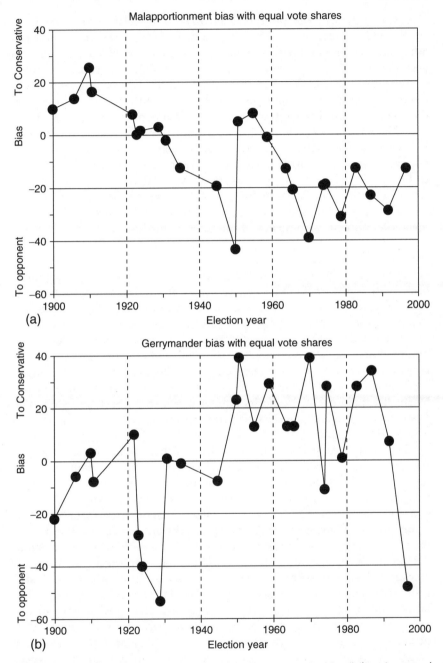

Figure 3.6. (a) Malapportionment and (b) gerrymander bias components at twentieth-century general elections (excluding 1918)

As noted earlier, malapportionment may be at least partly built in to an electoral system, if the average size of constituencies varies between different areas and one party is stronger in the areas with smaller constituencies. The 1944 Act guaranteed Scotland and Wales a minimum number of constituencies, and the 1958 amendment ensured that this meant – given the relative patterns of population change between those two countries, on the one hand, and England, on the other – that constituencies in Scotland and Wales would be relatively smaller than those in England (Rossiter, Johnston and Pattie 1999). Labour has increasingly outvoted the Conservatives in those two countries since 1950, and this has produced a bias of up to 14 seats in its favour, out of the total shown in figure 3.6a. The Conservatives lost out to Labour both through the 'over-representation' of Scotland and Wales in the House of Commons and through the process of creeping malapportionment within each of the three countries.

The gerrymander component of electoral bias is a consequence of different patterns of surplus and wasted votes between the two main parties: the smaller the number of each that a party gets, the better its performance, because its vote distribution is more efficient. In the first half of the century, the Conservatives were generally disadvantaged by this – especially in the 1920s – but for much of the second half the situation was reversed (figure 3.6b). The work of the Boundary Commissions involved the production of unintentional gerrymanders, which Gudgin and Taylor (1979) showed tend to favour Labour in the large cities and the Conservatives elsewhere, with the latter being the net beneficiary, largely because its candidates won by smaller majorities than did its Labour opponents where they won. The 1997 election was a major deviation from this pattern, however – a point to which we return below.

There is a third set of sources of electoral bias, what we have termed *reactive malapportionment* (Johnston et al. 2001). These result from the impact of other aspects of voting behaviour. Two of them in effect reduce the number of votes one of the main parties needs to win a seat – and therefore operate in the same way as classical malapportionment. The first is the number of abstentions: the more of these there are in a constituency, the fewer votes are needed to win there, so that a party which is strong in areas with high abstention levels is advantaged (according to our bias measure) relative to one which is strongest where turnout is high. And if turnout falls unevenly – as has been the case in recent decades – then the party which is strongest where it falls most will reap even greater advantages. The second operates in exactly the same way; the larger the number of votes for a minor party, the smaller the number of votes needed for victory by one of the other parties, so that the one which is strongest where a minor party or parties perform best will reap an advantage – unless a

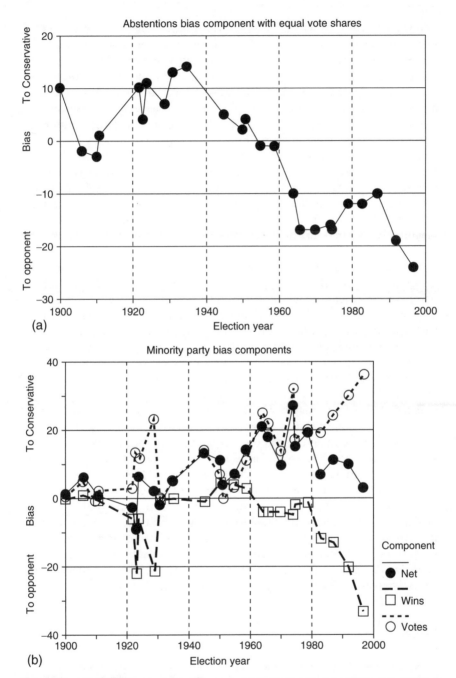

Figure 3.7. (a) Abstentions and (b) minor party bias components at twentieth-century general elections (excluding 1918)

minor party wins, in which case this will disadvantage the stronger of the two main parties (the third reactive malapportionment component).

The geography of abstentions largely favoured the Conservatives in the first half of the century – due to lower turnout rates in rural areas. But over the last five decades, the abstentions component produced an anti-Conservative bias (figure 3.7a): turnout was consistently lower in Labour-than in Conservative-held seats. Furthermore, as turnout nationally declined from 84.1 per cent in Great Britain in 1950 to 71.4 per cent in 1997, so Labour's advantage increased: exceeding 10 seats for the first time in 1964, and reaching 25 by 1997. Many Labour-held seats were very safe for the party: there was little incentive for individuals (especially those not very interested in politics) to go to the polling booth, and increasingly Labour did little to mobilize their support, spending much less on campaigning in its safe seats than the Conservatives did in theirs (Denver and Hands 1997; Johnston et al. 2001).

The disadvantage that the Conservatives suffered from the abstentions component of reactive malapportionment was countered by the minor-party votes element, which operated in its favour virtually throughout the century (figure 3.7b). There was an early peak in the 1920s, when Labour was replacing the Liberals as the second party and the two parties split the anti-Conservative vote in many constituencies, making them relatively easy for the Tories to win. It then increased rapidly from the 1960s on, as the Liberal revival (plus, to a lesser extent, that of the Scottish National Party (SNP) and Plaid Cymru in the non-industrial areas of Scotland and Wales) reduced the average majority needed to win in many Conservative-held seats more so than it did in those held by Labour; the Liberals and their allies became the main opposition party in much of the Conservative-held south of England (Johnston, Pattie and Allsopp 1988). However, when those minor parties won seats (in the 1920s and the 1980s–1990s), most of them would otherwise have been won by the Conservatives, which was a bias advantage for Labour (the wins component in figure 3.7b). The net pro-Conservative bias from these two minor-party components combined was largest in the 1970s (figure 3.7b), when the minor parties won many votes (up to one-quarter of the total) but few seats (never more than 10 per cent); as the minor parties won more seats, so the net advantage to the Conservatives declined, to virtually zero by 1997.

Conclusions

> No other party [than the British Conservatives] in a Western democracy can rival its electoral and campaigning record. (Davies 1995:196)

But to what extent can that record – no longer sustainable after the comprehensive defeats of 1997 and 2001 – be linked to the operation of the British electoral system? And did the Conservatives manipulate that system to their own ends, or were any benefits gained serendipitous?

In the first part of the twentieth century the system largely worked to the Conservatives' advantage. They were strongest in the areas with the smallest constituencies, had their support more effectively distributed throughout the country (with no major concentrations of surplus or wasted votes), and benefited when the Liberal and Labour parties split the opposition vote during the 1920s. There was no need for the Conservatives to do anything to the electoral system – and they didn't, with the 1918 redistribution being implemented by a Liberal-led coalition. (Although the Conservatives did consider shifting a substantial number of seats from Ireland to England in 1905 to bolster their cause.) The system put in place by Lord Salisbury in the nineteenth century continued to serve them well, despite the extension of the franchise – but whether this lack of concern to change the pattern of constituencies was intentional or occurred by default is difficult to tell.

During the 1930s, the Conservative government declined to introduce a redistribution which may have aided Labour, since the largest constituencies were by then in the urban areas. But when a number of important electoral issues needed resolution ready for the first election after the Second World War, the Conservative-led coalition acquiesced in the introduction of a regular series of independent redistributions, as the result of an independent inquiry chaired by the Registrar General. This was followed by a Speaker's Conference chaired by an erstwhile Conservative MP, which included a number of members who had been promoting redistribution in the 1930s, and then a Bill which encountered little opposition in the House of Commons. The party may have felt that it would be the main beneficiary of such regular reviews, because they would remove the creeping malapportionment which, with urban depopulation, was likely to benefit the Labour party over time. Indeed, it became part of the conventional wisdom during the 1950s–1980s that the Conservatives usually benefited by up to 20 seats from any redistribution, and it was this belief that led the party to bring forward the next redistribution in 1992 to ensure that new constituencies were available before the next general election (Rossiter, Johnston and Pattie 1999).

What the Conservatives didn't realize, however, was that this apparently inbuilt benefit from redistribution could be countered by detailed presentation of pro-Labour cases during the public consultation process – which the Conservative government made much more important in its 1958 amendments. In the 'battle of the boundaries' during the Fourth Periodic

Review (1991–5) Labour very successfully outflanked the Conservatives in several parts of the country, effectively achieving gerrymanders there by convincing the Commissioners of the validity of their alternative configurations for some constituencies. Furthermore, in the late 1980s Labour realized that once a set of constituencies had been established, careful targeting of voter mobilization can influence the outcome very substantially. There was little point in pressing for high levels of turnout in safe Labour seats, merely to garner more surplus votes; nor was there much point in pressing Labour supporters to go to the polls in safe Conservative seats, only to accumulate a few more wasted votes. Focused campaigning in marginal seats, perhaps assisted by anti-Conservative tactical voting in constituencies where Liberal Democrats had the best chances, was the basis of Labour's strategy in the 1990s (and 2001), whereas the Conservatives continued to activate their formidable electoral machine (aided by their greater financial resources) in places where more votes would not be effective. As a result, Labour was able to achieve a major change in the direction and volume of the gerrymander bias component in the 1990s, as well as gaining from the lower turnout (because many of the increased abstentions were in safe Labour seats).

In general, therefore, for much of the century the Conservatives took little interest in the details of the operation of the electoral system bequeathed them by Lord Salisbury. They were complacent, assuming that the major change of regular redistributions introduced in the 1940s would benefit them, and not being aware that carefully crafted strategies of operating both on and within those redistributions could work to their opponents' benefit. That complacency even included the guarantee of a minimum number of seats for Scotland and Wales, introduced in 1944 and in effect extended in 1958, which was then declared a 'political no-go area' in the 1980s (Rossiter, Johnston and Pattie 1999). Wales was already a very strongly pro-Labour area by the 1950s, Labour winning over 50 per cent of the votes cast there at the majority of elections post-1945; Scotland did not become so until the 1970s. Granting and then sustaining over-representation for Scotland and Wales ceded an electoral advantage to Labour which became enhanced over subsequent decades.

So, the Conservative century was assisted by the operations of the electoral system, but without the party doing much to enhance that. Largely through complacency, it seems, the Conservatives assumed that they would benefit from the system's quirks – and for much of the time they did. But when, in the last decade, their opponents pierced that bubble of complacency, the Conservatives in 1997 and 2001 suffered two of their worst election defeats ever. They realized, after Labour had proved the point beyond doubt, that geography matters in the conduct of British elections.

NOTES

1   In 1906 the Conservative party won 16 seats in Ireland, which returned 101 MPs then. If just Great Britain is considered, then the party won 25 per cent of the seats.

2   In 1923 the Conservative party got 38.1 per cent of the votes cast but 42 per cent of the seats.

3   Such information is never available to the parties, although the major competitors develop databases from their canvassing which identify (to the extent that voters will tell them) where their supporters live. Official records provide no data on voting patterns below the level of the constituency for parliamentary elections and the ward for local government elections.

4   There was also a small number of electors who, until 1949, had two – and in some cases three – votes: as well as the constituency in which they resided they could also vote for a representative of the university from which they graduated and/or as a member of the business roll.

5   As in all of the calculations reported here, these figures relate only to the constituencies contested by both of the main parties at the relevant election, since it is very difficult – especially for elections early in the century – to estimate what the number of votes won by the various parties would have been if they had been contested. The number of uncontested seats declined during the first half of the century. There were over 150 uncontested seats at the 1900 election, but this fell to 32 in 1906 and just 9 in the first (January) 1910 contest. It was 93 at the second (December) 1910 election but then less than 50 at each of the four contests in the 1920s and the two in the 1930s. There were just two uncontested seats in 1945 and none from 1950–1951 on (save those being contested by the incumbent Speaker). It is our guess – but no more – that a majority of the uncontested seats in the early part of the century were both small and Conservative strongholds.

REFERENCES

Abrams, M., R. Rose, and R. Hinden 1960: *Must Labour Lose?* London: Penguin Books.

Budge, I. 1999: Party ideology and policy: reversing the 1950s? In G. Evans and P. Norris (eds), *Critical Elections: British Parties and Voters in Long-Term Perspective*. London: Sage Publications.

Champion, A. G. 1998: *Counterurbanization: the Changing Pace and Nature of Population Change*. London: Edward Arnold.

Crewe, I. 1986: On the death and resurrection of class voting: some comments on *How Britain Votes*. *Political Studies*, 34, 620–38.

Crewe, I., and A. King 1995: *The SDP: The Birth, Life and Death of the Social Democratic Party*. Oxford: Oxford University Press.

Davies, A. J. 1995: *We, the Nation: The Conservative Party and the Pursuit of Power*. London: Little, Brown.

Denver, D. T., and G. Hands 1997: *Modern Constituency Electioneering: The 1992 General Election*. London: Frank Cass.

Dorling, D., C. J. Pattie, R. J. Johnston and D. J. Rossiter 1996: Missing voters in Britain 1992–1996: where, and with what impact? In D. Farrell, D. Broughton, D. Denver and J. Fisher (eds), *British Elections and Parties Yearbook 1996*. London: Frank Cass, 37–49.

Gudgin, G., and P. J. Taylor 1979: *Seats, Votes and the Spatial Organisation of Elections*. London: Pion.

Johnston, R. J., C. J. Pattie and J. G. Allsopp 1988: *A Nation Dividing? Britain's Changing Electoral Map, 1979–1987*. London: Longman.

Johnston, R. J., C. J. Pattie, D. F. L. Dorling and D. J. Rossiter 2001: *From Votes to Seats: The Operation of the UK Electoral System since 1945*. Manchester: Manchester University Press.

Johnston, R. J., C. J. Pattie, D. F. L. Dorling and D. J. Rossiter 2002: Distortion magnified: New Labour and the British electoral system, 1950–2001. In L. Bennie, C. Rallings, J. Tonge and P. Webb (eds), *British Elections and Parties Review*, vol. 12. London: Frank Cass, 133–55.

Johnston, R. J., and D. J. Rossiter 1982: Constituency building, political representation and electoral bias in urban England. In D. T. Herbert and R. J. Johnston (eds), *Geography and the Urban Environment*, vol. 5. Chichester: John Wiley, 113–54.

McKenzie, R. T., and A. Silver 1968: *Angels in Marble: Working-Class Conservatives in Urban England*. London: Heinemann.

Nordlinger, E. A. 1967: *The Working Class Tories: Authority, Deference and Stable Democracy*. London: McGibbon & Kee.

Pattie, C. J., D. F. L. Dorling, R. J. Johnston and D. J. Rossiter 1996: Electoral registration, population mobility and the democratic franchise: the geography of postal voters, overseas voters and missing voters in Great Britain. *International Journal of Population Geography*, 2, 239–60.

Pattie, C. J., and R. J. Johnston 1996: The Conservative Party and the electorate. In S. Ludlam and M. J. Smith (eds), *Contemporary British Conservatism*. London: Macmillan, 37–62.

Pinto-Duschinsky, M. 1981: *British Political Finance, 1830–1980*. Washington, DC: American Enterprise Institute for Public Policy Research.

Roberts, A. 1999: *Salisbury: Victorian Titan*. London: Weidenfeld & Nicolson.

Rossiter, D. J., R. J. Johnston and C. J. Pattie 1999: *The Boundary Commissions: Redrawing the UK's Map of Parliamentary Constituencies*. Manchester: Manchester University Press.

Salisbury, Lord 1884: The value of redistribution: a note on electoral statistics. *The National Review*, 4, 145–62.

Samuels, D., and R. Snyder 2001: The value of a vote: malapportionment in comparative perspective. *British Journal of Political Science*, 31, 651–72.

Seldon, A., and S. Ball 1995: Introduction. In A. Seldon and S. Ball (eds), *The Conservative Century: The Conservative Party since 1900*. Oxford: Oxford University Press.

Taylor, P. J., and G. Gudgin 1975: A fresh look at the Parliamentary Boundary Commission. *Parliamentary Affairs*, 28, 405–15.

Chapter Four

Mobility in the Twentieth Century: Substituting Commuting for Migration?

Colin G. Pooley

Introduction

The twentieth century has witnessed unparalleled changes in personal mobility due to new transport technologies, increased standards of living, shorter and more flexible working hours and broadened personal horizons. These, and other factors, have combined to enable many people to adjust housing needs to suit changing family circumstances, to commute to work over long distances, and to travel extensively for leisure and pleasure. Mobility is now so embedded in modern society that Urry (1999) has argued that mobility has reconstituted social life, and should replace society as the focus for sociological study in the twenty-first century. Such changes are usually perceived to be beneficial, increasing personal freedoms and widening opportunities; but high levels of mobility are also recognized to have social costs in terms of congestion, resource use and environmental damage (Schafer and Victor 1997; Whitelegg 1997; Adams 1999; Banister et. al. 2000). This chapter provides a brief historical overview of mobility trends in twentieth-century Britain, focusing especially on the changing relationship between residential location and the journey to work. Specifically, the chapter examines the ways in which social, economic, cultural and technological changes have enabled people to move house but not their job, and to seek new employment without moving home. It also considers the implications of such trends for twentieth-century society.

It is first necessary to examine the relationship between mobility and 'geographies of modernity'. Gregory (2000) highlights two axes of debate with respect to the construction of modernity. First, that modernity represents progress rather than tradition and, second, that it represents an acceleration of social change and a new consciousness of time. It can be argued that changed patterns, processes and perceptions of mobility are central to such transformations. In Britain, innovations in transport technology that allowed new levels of mobility occurred principally in the nineteenth century, with the development of the railway, the electric tram and the internal combustion engine (Dyos and Aldcroft 1969; Freeman and Aldcroft 1988). With the exception of air travel, which in a small country such as Britain has remained of peripheral importance for internal mobility for most people, progress in the twentieth century has been represented by the refinement of nineteenth-century technologies, and through their spread to a larger proportion of the population, rather than by the invention of new modes of transport. However, progress to modernity has been achieved both through the speeding up of transport producing the well-established process of time-space convergence (Pred 1973; Thrift 1990) and, most crucially, in the ways in which the population use and value time (Urry 1999). Instant access to news and online information have created a society in which, arguably, the traditional 'disciplines of time' (Thompson 1967) have broken down. An almost universal desire to fit more activities (work, family, leisure) into a finite amount of time has meant that life has speeded up in a way that demands high levels of mobility and new mobility experiences. It has been argued that this process of time-space compression has led to increased societal dislocation (Harvey 1989, 1990), but it can also be suggested that the implications of such changes are complex and may impact on different individuals in a variety of ways (Gregory 1994). Moreover, where new mobility experiences are not delivered, for instance through increased urban traffic congestion and the impact of under-investment in rail infrastructure in the late twentieth century, it has led to increasing frustration, with mobility expectations remaining unmet (Whitelegg 1997; Banister et. al. 2000). Differential access to transport and constraints on personal time, influenced by factors such as gender and income, further affect the extent to which individuals can meet their mobility expectations and benefit from new transport structures (Massey 1993; Hjorthol 2000). Over the twentieth century, time-space budgets (Carlstein, Parkes and Thrift 1978; Parkes and Thrift 1980) have become both more complex and increasingly significant in everyday life.

To understand fully the implications of these relationships, and especially of the contribution of mobility to twentieth-century modernity, it is necessary both to examine the nature of mobility and to think spatially. If it is argued that it is not mobility *per se* that is important, but rather the things that enhanced mobility allows people to do, it can be suggested that the transformation of time in the process of modernity can be achieved through a variety of different routes. Often this has required the substitution of one form of mobility for another. New computer technologies and the spread of the internet, for example, have allowed people to remain in one location (either at home or at work), and to use time in new ways to access information that previously would have required a degree of spatial mobility (Thrift 1995). Similarly, new and more complex patterns of mobility have allowed people to fit more activities into a given time period: rather than making separate trips to travel to work, shop or for leisure people increasingly combine activities so that a journey home from work may include a visit to a supermarket, a gym, or an after-school club to collect a child, thus producing more complex modern patterns of mobility. Furthermore, the ability to commute over long distances has meant that people can change work without moving home or, conversely, can move home without changing their job, thus substituting increased daily mobility for residential migration. It is this aspect of mobility transformation that is directly examined in this chapter. However, all such changes to some degree reconfigured the spatial structure of twentieth-century society and economy. None has necessarily reduced total demands for mobility (Black 2001), but it is these spatial transformations that inextricably link geography and modernity in twentieth-century Britain.

For the most part migration and mobility have been poorly theorized in the academic literature (Woods 1985; Findlay and Graham 1991; Champion and Fielding 1992; Courgeau 1995). Wilbur Zelinsky's hypothesis of 'mobility transition' suggests a framework within which assumed increased levels of mobility may be linked to the process of modernization (Zelinsky 1971), but both the hypothesis itself and the underlying assumptions about modernization have been heavily criticized (Woods 1993; Hoerder and Moch 1996; Pooley and Turnbull 1998; Hochstadt 1999). The theme of migration and modernity is tackled most directly by Hochstadt in his study of migration in Germany, 1820–1989. He combines detailed quantitative analysis of aggregate historical migration data for Germany with discussion of a theoretical framework for the study of mobility during the transition to modernity. Hochstadt uses German data to challenge the notion that levels of mobility necessarily increased during the process of modernization in the way that Zelinsky hypothesized. However, despite high-quality empirical evidence, he fails to develop a convincing alternative theoretical framework. Following from such critiques of migration theory, this chapter argues that any mobility transition during the twentieth century related more to changes in the nature and balance of mobility (which was indeed part of Zelinsky's argument), and to the ways in which mobility was used, rather than simply to increases in the amount of mobility. It also argues that an understanding of the nature of mobility substitution in the twentieth century can only be achieved by focusing on the individual experiences of migration and mobility, constructed over the life course, rather than through the analysis of aggregate data.

Data

There are no readily available British data that allow detailed comparative analyses of either migration or daily mobility over the twentieth century. Census data on migration fail to register all moves and, in some instances, only record movement between administrative units. Moreover, the ways in which the census recorded migration changed markedly during the century. Although recent censuses provide some data on the journey to work, and a range of surveys (most notably the British Travel Survey) provide some data on daily mobility, broadly comparative data are only available for the second half of the century. Information must thus be pieced together from a variety of sources, or collected retrospectively from individual respondents, and this chapter is based on two large data sets collected as part of ongoing research on migration and mobility in Britain since the eighteenth century. Details of the data collection and methods of analysis are presented elsewhere (Pooley and Turnbull 1998, 1999; 2000), and only the barest outlines are given here. The research adopted a biographical approach which focused on both residential migration and aspects of daily mobility. In the study of migration a total of 16,091 life histories was collected for people born between 1750 and 1930, of which 3,791 pertain to people born after 1880, and thus in employment during the twentieth century, and therefore relevant here. In a separate study, 1,834 detailed journey-to-work life histories were collected for people who entered the workforce between 1890 and 1980. In addition 90 in-depth interviews were conducted with people who entered the workforce from the 1920s. There are many potential biases in these data sets (see below), but together the data do begin to give a picture of changes in the relationship between migration and the journey to work during the twentieth century.

The data on residential migration contained detailed information on the migration life history of each individual together with data on occupations, family structure, reasons for migration and a range of associated variables. The information was collected through a network of family historians, contacted through their family history societies, who agreed to complete a data-entry form for the ancestors whose lives they had traced. Using this large volume of completed research, we constructed an elaborate data-collection mechanism to utilize work completed by family historians. We asked respondents to indicate the sources they had used to collect the information they provided, and this included a wide range of material, with family papers and diaries supplementing the more standard census and registration sources. Data were collected from all parts of the country and spanned the full time period under study.

All historical sources contain some biases and errors and this data set is no exception (Pooley and Turnbull 1998: 38–50). There are three main sources of error that may affect the analysis presented here. First, family historians, like other researchers, are limited by the sources available – with, for instance, much more information on reasons for migration in the twentieth century when oral history can be used than in the eighteenth century – and may have made errors in interpreting and recording data they have collected. Secondly, and more seriously, family historians may have introduced bias through their own selection of which ancestors to trace and a personal desire to 'tell a story' about their past. Thirdly, further bias may be introduced through our recording, analysis and interpretation of the information provided. All data received were carefully checked and any forms containing incomplete or inconsistent data were rejected. The aggregate characteristics of the data set at appropriate dates were compared with available data from the 1801, 1851, 1891 and 1951 censuses. This showed that the data contained some bias towards males, those who married at some time in their life, those who lived to an old age, and those in skilled occupations. This is not surprising. Men are easier to trace through historical sources than women who changed their name on marriage, and some family historians are principally interested in tracing a name down the male side of the family. By definition, all immediate ancestors of present-day family historians lived to adulthood and had children (family lines that died out are rarely traced), and the poor and unskilled are least likely to be traced through historical records. However, none of these biases is severe, and in other respects (for instance geographical distribution) the data offer a good representation of the total population. We conclude that the family historians whose data we used conducted their research with a high degree of accuracy and most of the biases are an inevitable result of the sources available. To the extent that they introduced particular biases due to their interests in family history, these are consistent and predictable. In all other respects the data form a large and accurate information source that, if interpreted with care, can provide valuable insights into past migration processes. In the context of this chapter it is also important to note that data for the twentieth century are likely to contain fewer biases than those for earlier periods.

Journey-to-work data were collected for people who began work in each decade from the 1890s to the 1980s. Information on those who began work in the late nineteenth and early twentieth centuries was collected through the network of family historians, who were asked to provide information on their immediate ancestors (usually their parents). For those who began work from about the 1930s, respondents were identified in a number of ways. Family historians were asked to provide data for themselves, and respondents were contacted through a series of large employers, through local newspapers, other media, and pensioners' clubs. All respondents were sent a data-entry form requiring information on an individual's employment history, details of every new journey to work undertaken, residential history and personal characteristics. The largest number of responses related to people who began work in the mid-twentieth century: 1,040 (56.7 per cent) began work in the four decades 1930–70. The characteristics of respondents were compared with those recorded as being in employment in the 1931, 1951 and 1971 censuses of Great Britain. Overall, the sample is quite representative of the total population with respect to gender, age and marital status, though with some under-representation of women in the 1930s, and of young and single workers in the 1970s, but does contain a substantial bias towards those in higher socio-economic groups, particularly in the 1970s. All those who provided life-history data relating to themselves were asked if they would be prepared to participate in an in-depth interview, probing their experience of the journey to work in the past, and focusing especially on their reasons for making particular decisions about modes of transport and the location of home and workplace. A total of 90 respondents were selected for interview: 50 who began work in the 1920s, 1930s and 1940s were interviewed face to face, each interview lasting 1.5–2 hours, and a further 40 who began work later in the century were interviewed more briefly by telephone. All interviews were taped, transcribed and analysed using the computer text analysis program NU.DIST. All those interviewed had begun work in one of three labour markets: London, Manchester or Glasgow, and they were selected to provide a balance of male and female respondents and to provide a broadly representative socio-economic cross-section of the population. Thus those in unskilled work were over-sampled in relation to their presence in the life history survey data.

Changing Home and Workplace

British society has been characterized by high levels of short-distance residential mobility since at least the eighteenth century, with substantial amounts of movement both up and down the urban hierarchy, within towns, between towns and within rural areas at all time periods (Pooley and Turnbull 1998). However, there is some evidence that the frequency of residential movement increased slightly in the twentieth century, though long-run historical data must be interpreted with caution as many short-distance moves are likely to be under-represented in sources for earlier time periods. From the sample of residential life histories, people born 1890–1900 moved on average 6.6 times during their lifetime, whereas those born in the 1920s moved on 8.1 occasions. Data from the 1,834 life histories recording housing, employment and journey-to-work change show that the number of residential moves undertaken during a working life increased modestly during the twentieth century, from 4.5 moves for men entering the labour market in the 1890s, to over 5 during a shorter working life for those starting work in the 1950s (table 4.1). Information on those entering the labour market more recently cannot be directly compared since, although there is some evidence that levels of mobility have increased more rapidly in the 1970s and 1980s, it is likely that most mobility occurs in the early stages of the life course. However, Rees (1979) estimated that in the 1970s the average Briton moved between 7 and 11 times during their lifetime, with most such moves consisting of short-distance intra-regional relocation (Coleman and Salt 1992). The data also indicate the convergence of male and female employment and mobility patterns. Whereas in the early twentieth century females moved home less frequently than males, from the 1970s female residential mobility matched and then exceeded that of males.

Although most moves in all time periods were over relatively short distances, the late twentieth century was also characterized by an increase in both the number and distance of long-distance moves (Owen and Green 1992). Looking, first, at long-run trends in residential migration, life-history data show that, whereas the mean distance moved 1880–1919 was 38.4 km (a figure that had remained more or less constant since the mid-eighteenth century), for the period 1920–94 this increased to 55.5 km. Similarly, while only 12.4 per cent of moves were over 100 km in

Table 4.1. Mean number of residential moves undertaken during the working life, by gender and decade of entering the labour market

	Mean number of moves		Mean number of years in labour market	
Decade	Males	Females	Males	Females
1890–99	4.5	2.5	46.4	28.8
1900–9	4.5	2.7	46.9	29.0
1910–19	4.7	2.9	47.3	33.4
1920–29	5.0	3.3	44.3	30.7
1930–39	5.4	4.2	44.8	35.8
1940–49	5.5	4.1	43.2	33.8
1950–59	5.0	4.5	35.2	38.5
1960–69	5.3	4.7	33.4	30.6
1970–79	4.9	4.6	22.8	21.5
1980–89	3.1	4.2	13.2	13.1

Source: Calculated from 1,834 individual life histories.

1880–1919, this increased to 18.7 per cent of moves in 1920–94 (Pooley and Turnbull 1998: 65). These changes were also reflected in the spatial patterning of residential moves. In 1880–1919, 47.9 per cent of moves took place within the same settlement, but in 1920–94 this fell to 37.3 per cent as more residential mobility occurred between settlements. There were also small changes in the relative attractiveness of different types of settlement. The proportion of moves to a new settlement within the same size band remained quite constant (around 16 per cent) from the 1880s, but in the period 1880–1919, whereas those moving up and down the urban hierarchy were evenly balanced (*c.*17 per cent of moves), after 1920 moves from large to small settlements accounted for 25.4 per cent of all mobility compared to 20.4 per cent of moves that went up the urban hierarchy (Pooley and Turnbull 1998: 97). Trends towards counter-urbanization, clearly identified in Britain from the 1970s (Champion 1989) were already emerging in the early twentieth century (Pooley and Turnbull 1996).

There have also been substantial changes in the average distance over which people travel to work and, especially, in the number of commuters travelling long distances (Pooley and Turnbull 1999). Whereas at the start of the century the mean commuting distance was less than 4 km, by the 1990s it had more than trebled to over 14 km (table 4.2). The same trends were apparent in all parts of Britain, but commuting journeys into London were consistently double those in other locations. Longer commuting journeys, made possible by new forms of transport, enable people to

Table 4.2. Average distance (km), and time (min.) travelled for journeys to work since 1890 by gender

	Males		Females		All	
Decade	Dist.	Time	Dist.	Time	Dist.	Time
1890–99	4.0	17.0	1.8	21.3	3.6	17.7
1900–9	3.9	21.5	3.2	25.4	3.8	22.4
1910–19	6.2	27.0	5.1	26.8	5.9	27.0
1920–29	6.8	28.2	6.1	31.3	6.7	29.0
1930–39	7.0	30.5	6.8	31.9	7.0	30.9
1940–49	8.2	33.8	7.3	33.1	7.8	33.5
1950–59	10.1	33.6	7.4	34.4	9.0	33.9
1960–69	12.1	34.6	7.5	32.1	10.2	33.5
1970–79	13.1	34.5	7.6	28.5	10.3	31.5
1980–89	15.5	37.3	8.8	29.4	12.0	33.1
1990–98	19.4	39.1	10.5	30.7	14.6	34.5

Source: Details of 12,439 journeys to work taken from 1,834 individual life histories. Statistics relate to all modes of transport and are calculated for the decade in which a particular journey to work started.

Table 4.3.   Main mode of transport for journeys to work since 1890 by gender (%)

| | *Time period* | | | | |
Mode	*1890–1919*	*1920–39*	*1940–59*	*1960–79*	*1980–98*
Males					
Walk	44.3	23.2	12.3	8.0	7.2
Bicycle	13.2	21.6	21.4	5.5	7.1
Tram/trolley bus	13.8	9.0	3.3	0.0	0.0
Bus	9.8	10.8	17.7	11.3	5.8
Train (overground)	14.4	17.5	18.9	17.9	21.9
Underground	1.2	3.2	3.8	3.9	4.0
Motor cycle	0.5	3.5	3.9	3.3	1.8
Car/van	1.6	10.0	16.3	48.1	50.9
Sample size	646	1,272	2,291	2,026	1,000
Females					
Walk	50.3	29.3	19.7	20.6	11.6
Bicycle	5.0	9.5	11.7	4.0	5.1
Tram/trolley bus	17.8	13.2	6.2	0.2	0.1
Bus	14.2	22.2	31.7	24.8	16.1
Train (overground)	8.6	20.3	18.1	11.1	10.1
Underground	3.6	3.9	6.6	7.2	6.1
Motor cycle	0.0	0.9	0.6	1.1	1.0
Car/van	0.5	0.7	4.4	29.9	49.0
Sample size	197	433	1,457	1,677	1,108

Source: Details of 12,439 journeys to work taken from 1,834 individual life histories. Statistics are calculated for the decade in which a particular journey to work started.

commute over longer distances without necessarily increasing the time committed to travelling to and from work. Thus the average amount of time spent commuting one way to work barely doubled over the twentieth century, with most of that increase occurring before the 1920s. From the 1930s to the 1990s the mean time spent commuting increased by less than five minutes, from 30.9 minutes to 34.5 minutes. As with distances travelled, journey times have been consistently greater in London than in other locations (table 4.2). The modal shifts that have allowed commuters to travel further without increasing travel times are clearly shown in table 4.3. Whereas the use of all forms of public transport has remained fairly constant, commuting on foot and by bicycle has declined steeply, and the use of motor vehicles has increased rapidly. There are also distinct gender differences: at all time periods women were more likely than men to use public transport and less likely to cycle to work; and during the second half of the twentieth century women were slower than men to switch to commuting by car.

These data suggest that most people have a threshold of travel time that they are willing to spend commuting to work, and that the distance they travel is determined largely by the transport technology available to them. It can also be suggested that the ability to commute over longer distances, without significantly increasing travel times, has given people the opportunity to adjust their mobility behaviour by substituting commuting for residential migration. Green, Hogarth and Shackleton (1999) identify this as a recent trend, but life-history data suggest that the same processes have been operating for much of the twentieth century. Thus in 1890–1919, 59.1 per cent of people who moved home also changed their workplace; but in the period 1980–98 only 23.7 per cent of respondents who moved home also changed where they worked (table 4.4). Conversely, whereas in 1890–1919 only 20.8 per cent of people changed their workplace without moving home, in the 1980s and 1990s 43.8 per cent changed their job but remained in the same house.

It can thus be suggested that the twentieth century saw the interaction of two rather contradictory mobility trends. On the one hand, people progressively moved home more frequently and over longer distances. On the other hand, the ability to commute over increasing distances in approximately the same time allowed people to either change their job without moving home,

Table 4.4. The changing relationship between home and workplace as expressed through the journey to work, 1890–1998 (%)

	Time period					
	1890–1919	1920–39	1940–59	1960–79	1980–98	1890–1998
Change of home but no change of workplace	18.2	31.6	23.2	23.2	25.1	24.4
Change of workplace but no change of home	20.8	23.7	30.0	37.4	43.8	33.1
Change of home and workplace	59.1	41.9	41.8	33.9	23.7	37.5
Change of journey mode or route, but no change of home or workplace.	1.9	2.8	5.0	5.5	7.4	5.0
Sample size	913	1,778	3,822	3,780	2,146	12,439

Source: Details of 12,439 journeys to work taken from 1,834 individual life histories. Statistics relate to all modes of transport and are calculated for the decade in which a particular journey to work started.

or to move house without changing their work. This leads to relatively less residential mobility for employment reasons, but more for housing, environmental and lifestyle reasons. This is borne out by migration life-history data where, in 1880–1919, 38.8 per cent of moves were for work reasons, 16.0 per cent for housing reasons and 2.9 per cent for family reasons; but in 1920–94 only 26.3 per cent of moves were for work reasons, with 23.6 per cent for housing reasons and 5.7 per cent for family reasons (Pooley and Turnbull 1998: 72). Whereas in the early twentieth century, when most people walked or cycled to work, a 30-minute journey restricted them to living within a few kilometres of their workplace, in the late twentieth century these restrictions were significantly reduced. These changes acted both to increase the potential for residential mobility (people can move freely and improve their housing status within an acceptable commuting radius of their work), and to reduce it (people can more easily change employment without moving home). The modest increase in mobility indicated by the life-history data is thus a compromise arising from these competing trends. Moreover, it can be suggested that the most important factor in this balance is the interaction between the length of time that people are prepared to spend commuting, the speed of available transport, and the strength of ties to a particular neighbourhood.

These themes can be illustrated through the individual life histories of commuters who were interviewed. Thus one female respondent, a London social worker in the 1950s, chose a long and unpleasant journey to work mainly because of a preference for living in a particular part of London:

> Yes, I always intended to [move] but I never got round to it. There weren't that many places to live around Walworth that you know, it was a very working-class area of London and there weren't little flats and bedsitters in that area, they were all in Earl's Court, Bayswater in those days and Blooms-bury, and there wasn't, I intended to move and never did, and then I got a boyfriend in Blackheath so there was even less incentive. (R03)

Later in life the same woman put up with a long and complicated journey to work because of uncertainty both about her job and other aspects of her life:

> Well I liked the flat really which I suppose is mainly it, and also I was beginning to think I'd had enough of social work and everything by then and I didn't know whether I was, it was all in the melting pot, and my mother was very ill which was one of the things, and I didn't quite know what was going to happen if I was going to be called back home to help out or whatever, and then in the middle of it all I met D and so that was all a possibility, and so I just went on for three years. (R03)

For this respondent, the ability to commute over relatively long distances enabled her to remain in the same location (for largely personal reasons), even though the journey to work was inconvenient. This option would not have been available to her half a century earlier.

Changing Attitudes to Mobility

Qualitative evidence also allows an examination of changing attitudes towards mobility and different forms of transport during the twentieth century. As outlined above, it can be suggested that changing expectations of mobility were as much part of the transition to modernity as any physical changes in the type, speed or availability of different forms of transport. Thus even in the 1950s cars were perceived as a mode of transport to be used mainly for leisure purposes rather than commuting to work:

> If you had a car at that stage, if you had access to a car you would have used that for leisure use only. It would not have occurred to you to use it for work. (R04)

But by the 1970s the need to make a complicated cross-city journey to work in Manchester meant that the private car became the only viable option for this respondent:

> That was easiest again. No transport you see and I didn't fancy cycling in those days. You get past the bicycle stage. About ten minutes in a car which is across/most of these journeys were across country you see where transport wasn't really available. It would be two buses again. (R04)

By the 1960s more than one third of all journeys to work were undertaken by car, compared to just 6 per cent in the 1940s, and in the twenty-first century the motor car has become the ubiquitous form of travel for all types of personal mobility. Not only does it allow easy movement over long distances, but also it is increasingly used for a majority of short trips within urban areas. Changed use of the car epitomizes the transformation of attitudes towards different forms of transport in the twentieth century. Whereas those forms of transport that dominated in the early twentieth century (walking, cycling, public transport) were often slow, public and exposed to the elements, the car is precisely the opposite. It is fast, enclosed, individualistic and independent of others. While travelling by car it is possible to create an artificial environment in which levels of heat, light and ventilation are adjusted to suit personal tastes, in-car entertainment can be selected, and yet long-distance communication with the outside world can be maintained through the use of mobile phones. The car becomes almost as comfortable as the home, and provides a working environment almost as convenient as the office. Whereas in the past long-distance travel was often a major undertaking and a notable experience, everything about the modern car is designed to make travel as much like other aspects of everyday life as possible. By the 1990s the car was no longer a luxury to be used for leisure purposes, but a necessity that met all the perceived requirements of modern mobility. This is illustrated by the testimonies of two respondents from Glasgow:

> I think once you start using it [the car] I find that's the trouble, the thing seduces you. (R76)
> I've used it [the car] for twenty odd years of having one you know...The... thing about public transport is the public...it's not always a pleasant experience sharing it with some folk you know. (R70)

Although on average people have gained more flexibility in where they live and work, and can undertake longer journeys to work to suit both residential and employment preferences, this trend has not been universal. Poverty or family commitments have meant that for some people, especially women, the tight ties between home and workplace that were common in the early twentieth century continued into mid-century. Thus one respondent recalled how both she and her husband changed their employment after being rehoused in Manchester in the 1960s because their new home was too far from their employment:

> A. wasn't earning a lot and I was only part time. We thought well I could do with getting a job nearer to home so. Anyway we thought about so – cut a long story short – A. did get a job at/as caretaker on the estate just about 10 minutes' walk from where we lived and then two or three months later I was able to join him at school as/oh initially I'd just started on as a Dinner Lady doing dinner duties but then they wanted a cleaner so I gave up the dinner duties and became cleaner so A. and I were caretaker and cleaner at this particular school (R19)

The same respondent stressed how, in the 1950s and 1960s, the need to work close to home to fulfil childcare and household responsibilities constrained her employment and journey to work:

> I would take J. to my mother's each morning, twenty minutes walk there, I'd leave J. at my mother's etc. then my mother would meet me as I came out of work so my mother had her walk home again and I had my twenty minute walk home again with J. ready to be in time to see to A.'s lunch when he came for his lunch-time break. (R19)

These accounts again emphasize that for most people travel time was more important than distance, and that in this respect individual conceptions of mobility changed little over the course of the twentieth century. For those with access to new transport technology the ties between home and workplace were loosened, allowing decisions about employment and residential location to be made relatively independently of each other. However, for those constrained by low income, family commitments or other factors the need to remain within about 30 minutes' travel time of the workplace continued to link decisions about residential location and workplace in ways that have changed little since the early nineteenth century.

Conclusions

The general mobility trends reported in this chapter are well known, but a focus on mobility substitution and, especially, the ways in which residential mobility and daily mobility have been traded against each other, is novel. Moreover, it is argued that the analysis of individual life histories allows a much more detailed examination of long-run historical trends than has previously been possible. The data on which this chapter is based are far from perfect; for instance it has not been possible to incorporate information on the changing costs of commuting, but they do allow some broad comparisons to be made about mobility and migration over the century. It is argued that, with respect to mobility, one distinctive feature of the geography of British modernity in the twentieth century was that people developed new ways of conceptualizing and utilizing mobility. Thus, in addition to the wider availability of faster forms of transport, people increasingly came to use and view mobility in different ways, and, in so doing, constituted new sets of relationships with the spaces through which they moved. This chapter has not been able to explore all the complex dimensions of these interactions, but it is suggested that one key feature of mobility change was the extent to which some people were increasingly able to trade residential migration and daily mobility against each other. If this is the case, it suggests that studies of changing migration patterns and processes cannot be understood without reference both to changes in other forms of mobility, and to the spatial consequences of such changes.

The changes identified in this chapter also have implications for contemporary transport policy and urban planning (DETR 1998a, 1998b, 1999). Policies that seek to restrict car use may begin to reverse the trends identified here, once again strengthening ties between home and workplace, and encouraging people to live either close to employment, or near to fast and efficient public transport routes. This may in turn influence patterns of housing demand in urban areas. The mobility trends of the twentieth century will only continue if most people continue to have access to increasingly fast and cheap means of personal travel, or if telecommunications enable more people to work from home. It is by no means certain that such trends will continue or, if they do, that the outcomes will be beneficial (Adams 1999; Black 2001).

ACKNOWLEDGEMENTS

The author would like to express his thanks to Dr Jean Turnbull, who was employed as Research Associate on both the projects that provide data used in this chapter, and to the Economic and Social Research Council and the Leverhulme Trust for funding. The project team is particularly indebted to all respondents who provided data for the research or who agreed to be interviewed.

REFERENCES

Adams, J. 1999: The social implications of hypermobility. In OECD, *Project on Environmentally Sustainable Transport: The Economic and Social Implications of Sustainable Transport*, Paris: ENV/EPOC/PPC/T(99)3/FINAL, 75–113.

Banister, D., D. Stead, P. Steen, J. Akerman, K. Dreborg, P. Nijkamp and R. Scleicher-Tappeser 2000: *European Transport Policy and Sustainable Development*. London: Spon.

Black, W. 2001: An unpopular essay on transportation. *Journal of Transport Geography*, 9, 1–11.

Carlstein, T., D. Parkes, and N. Thrift (eds) 1978: *Timing Space and Spacing Time*, vol. 2: *Human Activity and Time Geography*. London: Arnold.

Champion, A. (ed.) 1989: *Counterurbanization: The Changing Pace and Nature of Population Deconcentration*. London: Arnold.

Champion, A., and A. Fielding (eds) 1992: *Migration Processes and Patterns*, vol. 1: *Research Progress and Prospects*. London: Belhaven.

Coleman, D., and J. Salt 1992: *The British Population: Patterns, Trends and Processes*. Oxford: Oxford University Press.

Courgeau, D. 1995: Migration theories and behavioural models. *International Journal of Population Geography*, 1, 19–28.

DETR (Department of Environment, Transport and the Regions) 1998a: *A New Deal for Transport: Better for Everyone*. Government White Paper on the Future of Transport. London: Stationery Office.

DETR 1998b: *Breaking the Logjam*. Government Consultation Paper on Fighting Traffic Congestion and Pollution through Road User and Workplace Parking Charges. London: Stationery Office.

DETR 1999: *From Workhorse to Thoroughbred: A Better Role for Bus Travel*. London: Stationery Office.

Dyos, H. J., and D. Aldcroft 1969: *British Transport: An Economic Survey from the Seventeenth Century to the Twentieth*. Leicester: Leicester University Press.

Findlay, A., and E. Graham 1991: The challenge facing population geography. *Progress in Human Geography*, 15, 149–62.

Freeman, M., and D. Aldcroft (eds) 1988: *Transport in Victorian Britain*. Manchester: Manchester University Press.

Green, A., T. Hogarth and R. Shackleton 1999: Longer distance commuting as a substitute for migration in Britain: a review of trends, issues and implications. *International Journal of Population Geography*, 51, 49–67.

Gregory, D. 1994: *Geographical Imaginations*. Oxford: Blackwell.

Gregory, D. 2000: Modernity. In R. J. Johnston, D. Gregory, G. Pratt and M. Watts (eds), *Dictionary of Human Geography*. Oxford: Blackwell, 512–16.

Harvey, D. 1989: *The Condition of Postmodernity*. Oxford: Blackwell

Harvey, D. 1990: Between space and time: reflections on the geographical imagination. *Annals of the Association of American Geographers*, 80, 418–34.

Hjorthol, R. 2000: Same city – different options: an analysis of the work trips of married couples in the metropolitan area of Oslo. *Journal of Transport Geography*, 8, 213–20.

Hochstadt, S. 1999: *Mobility and Modernity: Migration in Germany 1820–1989*. Ann Arbor: University of Michigan Press.

Hoerder, D., and L. P. Moch (eds), 1996: *European Migrants: Global and Local Perspectives*. Boston: Northeastern University.

Massey, D. 1993: Power-geometry and a progressive sense of place. In J. Bird, B. Curtis, T. Putnam, G. Robertson and L. Tickner (eds), *Mapping the Futures: Local Cultures, Global Change*. London: Routledge, 59–69.

Owen, D., and A. Green 1992: Migration patterns and trends. In A. Champion and A. Fielding (eds), *Migration Patterns and Processes*, vol. 1: *Research Progress and Prospects*. London: Belhaven, 17–40.

Parkes, D., and N. Thrift 1980: *Times, Spaces and Places: A Chronogeographic Perspective*. New York: Wiley.

Pooley, C., and J. Turnbull 1996: Counterurbanization: the nineteenth-century origins of a late-twentieth century phenomenon. *Area*, 28, 514–24.

Pooley, C., and J. Turnbull 1998: *Migration and Mobility in Britain since the Eighteenth Century*. London: UCL Press.

Pooley, C., and J. Turnbull 1999: The journey to work: a century of change. *Area*, 31, 281–92.

Pooley, C., and J. Turnbull 2000: Modal choice and modal change: the journey to work in Britain since 1890. *Journal of Transport Geography*, 8, 11–24.

Pred, A. 1973: *Urban Growth and the Circulation of Information: The United States Systems of Cities 1790–1840*. Cambridge MA: MIT Press.

Rees, P. 1979: *Migration and Settlement (I): United Kingdom*. Luxembourg: HASA.

Schafer, A., and D. Victor 1997: The past and future of global mobility. *Scientific American*, October, 36–9.

Thompson, E. P. 1967: Time, work-discipline and industrial capitalism. *Past and Present*, 37, 56–97.

Thrift, N. 1990: Transport and communication 1730–1914. In R. Butlin and R. Dodgshon (eds), *An Historical Geography of England and Wales*. London: Academic Press, 453–86.

Thrift, N. 1995: A hyperactive world. In R. Johnston, P. Taylor and M. Watts (eds), *Geographies of Global Change: Remapping the World in the Late Twentieth Century*. Oxford: Blackwell, 18–35.

Urry, J. 1999: *Sociology Beyond Societies: Mobilities for the Twenty-First Century*. London: Routledge.

Whitelegg, J. 1997: *Critical Mass: Transport, Environment and Society in the Twenty-First Century*. London: Pluto.

Woods, R. 1985: Towards a general theory of migration. In P. White and P. van der Knap (eds), *Contemporary Studies of Migration*. Norwich: Geobooks, 1–5.

Woods, R. 1993: Classics in human geography revisited: commentary. *Progress in Human Geography*, 17, 213–15.

Zelinsky, W. 1971: The hypothesis of the mobility transition. *Geographical Review*, 61, 219–49.

Chapter Five

Qualifying the Evidence: Perceptions of Rural Change in Britain in the Second Half of the Twentieth Century

Alun Howkins

The history of agriculture in the twentieth century is, at one level, one of decline. In 1918 agriculture accounted for 6.5 per cent of GDP; in 2001 it fell for the first time below 1 per cent (Countryside Agency 2001: 6). The greatest decline has been since 1950, when agriculture accounted for 5.8 per cent (Holderness 1985: 172). Alongside this was a decline in the British agricultural population: in 1921 just over 900,000 men and women made a living from agriculture; in 1981 it was just above 300,000 – 1.3 per cent of the population, 0.3 per cent more than 'Literary, artistic and sports' occupations (*Population Trends* 1987: 41).

The effects of this change are enormous, but one concerns the changing cultural relationship between town and country in the second half of the century. As the countryside has declined in importance as a site of agricultural work it has increased in importance as the focus of other areas of human life. Physically it has become the site of leisure – of non-work – simply 'living' in, retiring to and visiting the country. Ideologically, social and political definitions of the countryside have seen conflicts in which the 'urban' idea of the rural has gained importance. While not new, the balance of forces here has changed, even if the 'country' lobby, particularly the Countryside Alliance, the National Farmers' Union and the Country Landowners' Association, refuses to accept this and continues to claim a privileged viewpoint. Despite all post-war governments' bipartisan approach to agricultural policy, what has become ever more important are the perceptions of a predominantly urban population concerning what the countryside was, is and should be. The analysis of these perceptions is therefore of paramount importance. What do people think has happened in rural areas since the Second World War and why? Has the change essentially been 'good' or 'bad'? What follows is an attempt at a qualitative account of post-war agricultural and rural change, and a discussion of how these views square with different historical interpretations. I will end with some brief remarks about 'policy'.

The main qualitative source here is a 1995 Mass-Observation Directive, issued to collect material on the contemporary countryside and attitudes towards living in it. A final question collected data on perceptions of countryside change:

> Finally, and perhaps a bit difficult, no matter how old you are – urban or rural – do you think the countryside has changed in your lifetime? (Or since the war? Or in this century?)[1]

Mass-Observation (M-O) was founded in 1937 to create an 'anthropology of ourselves', by 'observing' people in social situations, writing about public opinion, committing the observers' own thoughts to paper, and keeping diaries – the latter mainly during the Second World War. The organization ceased operation in 1948–9, but the archive came to the University of Sussex in the 1970s. In 1981 the current archivist, Dorothy Sheridan, restarted the organization, concentrating on diary-writing and on two 'directives' per year. Directives offer open-ended questions, encouraging those replying to write at length (Stanley 1981; Bloome et al. 1993). The observers are a self-selecting group – they are not a 'sample'. In fact, 249 women replied to the 1995 directive and 89 men. The only area where they correspond to a 'national' sample is in regional distribution, with both the M-O respondents' place of residence and the national population distribution similarly skewed to the south and east. Age is more problematic, the M-O group containing a disproportionate number over 50. However, given the historical interest of the directive this had some advantage. But in terms of class there are real problems, with M-O being self-selecting, primarily drawn from the 50+ age group and biased towards often retired white-collar workers. In a sense then we are discussing middle-aged, and middle-class opinion. This is less worrying than it might be in other circumstances, but the most obvious drawback is that this group is likely to have stronger opinions than the population as a whole and to be more articulate on 'class' grounds. Members of the group are also likely to be more moderate in their views given the 'conservative' bias introduced by age, class, and the fact that most live in prosperous areas of the south-east.

Before moving on, the posed 'question' should be re-examined. At one level this may also be problematic. The countryside, and particularly its recent past, is especially amenable to construction as a 'golden age' of an idyllic lost Eden (Williams 1973), a tendency shaping literary, artistic and political impulses since Roman times. Nevertheless, this does not mean that such a valuation – that the past was 'better' – is necessarily wrong. Some work, for instance Ian Dyck's study of Cobbett, has argued convincingly that views on the deterioration of aspects of rural life had a firm basis in fact (Dyck 1992). Similarly, Jeanette Neeson's study of the common land economy of the late eighteenth and early nineteenth centuries shows that those among the poor who lamented the loss of commons had more on their side than historians have tended to think (Neeson 1993). More importantly, it is possible to 'check' specific points made by the M-O group against other accounts of the period.

The history of agriculture since the Second World War is usually presented as a triumph. At the end of the war farming organizations and, more importantly, public opinion saw British agriculture as fundamental to victory. In his conclusion to the 'official' history of wartime agriculture Keith Murray wrote: 'this history should be, without question, a "success story" – successful far beyond the calculations and estimates of the pre-war planners' (Murray 1955: 340). Although later writers, notably John Martin and Paul Brassley, have qualified this picture, the overall achievement remains impressive (Martin 2000; Brassley 2000). In less than six years British farmers had increased output enormously. Any losses were mostly in 'luxury' products such as butter, cheese and above all eggs. Similarly there were relative drops in livestock production. However, in terms of national self-sufficiency the change covers these areas too. Between 1939 and 1945 British farmers' share of the market had increased by 9 per cent in wheat, 47 per cent in barley, 12 per cent in sugar, 13 per cent in beef and 9 per cent in bacon and ham (Holderness 1985: 174). Increases were gained by both increasing the area under production and intensifying the use of land. Most obvious physically was the transformation of landscape through the ploughing up campaign. In 1941 a Sussex schoolteacher wrote in her diary about the appearance of great areas of 'brown' as opposed to green as they ploughed up the Sussex Weald with 'War Ag' tractors.[2] Nationally, by the end of the war this process reduced the area of permanent pasture by 6 million acres. Very little would return to pasture after 1945. The change to more intensive farming is less easy to summarize. Many gains were made by increased efficiency in manpower and existing resources, and increased use of machinery. More worrying for the long term was the beginning of widespread use of inorganic fertilizers and pesticides. In 1939–40 the government paid out £9,000 per annum in subsidy for the control of pests; in 1944–5 it paid £52,000. Further, as Murray points out the war saw a large increase in 'the control of weeds by selective weed killers' (Murray 1955: 384, 260).

These changes, but also the high opinion in which British farming was held during the war, were apparent in replies to an M-O directive of September 1942 asking 'how the problems of agriculture should be dealt with after the war'.[3] In general replies were very positive on farmers' response to the war,

Table 5.1.   Index numbers of agricultural output in the UK 1940–1945 (1936–7 to 1938–9 = 100). Crops

Crop	1940	1941	1942	1943	1944	1945
Cereals	132	155	182	195	186	179
Potatoes	131	164	193	202	187	201
Sugar Beet	116	118	143	137	119	141
Vegetables	110	122	156	133	144	137
Fodder crops	98	127	188	135	140	132

Table 5.2.   Index numbers of agricultural output in the UK 1940–1945 (1936–7 to 1938–9 = 100) Livestock. (Year June–May)

	1940–1	1941–2	1942–3	1943–4	1944–5	1945–6
Milk	90	88	93	96	97	100
Beef and veal	97	73	83	83	92	93
Mutton and lamb	108	89	89	79	72	69
Pigmeat	87	38	35	32	35	38
Eggs	90	75	57	51	54	63

Source: Brown 1987: 130

with a clear sense of harsh pre-war agricultural conditions. A Hertfordshire woman said: 'the farmers have done their damnedest for the country, it should be seen that they are not allowed to go back to penury and difficulty'.[4] A Preston man wrote: 'agriculture must never be allowed to fall into the condition into which it had descended prior to the war in this country'.[5] However, although the vast majority were 'pro-farmer' a large group were critical of aspects of wartime farming in ways which suggest that public opinion, or sections of it, was more forward-looking than many of those in government. A clerk from Oswestry in Shropshire wrote a lengthy critique of changes in wartime agriculture ending:

> I do hope that our farmers first, and then our authorities and people in general will learn the folly of treating the soil as a medium for growing things in. Such a conception leads to the belief that all we have to do is to discover by laboratory methods what each crop needs for its growth and then to see to it, merely by adding chemicals to the soil.[6]

Nor was the Oswestry respondent alone in this view, suggesting that among the M-O group at least some wartime and pre-war ideas of the 'humus' farming campaign had made an impact. As Phillip Conford has suggested wartime pamphlets and radio appearances by 'pro-organic' agriculturalists, especially Sir George Stapledon, may have advanced the organic criticism of modern agriculture (Conford 2001). A man from Brixton urged that farmers should adopt, 'the system of ley farming advocated by Prof. Stapledon...'.[7]

There was widespread worry about urban and suburban encroachment, again suggesting that pre-war debates had made an impact. A London woman wrote: 'a comprehensive scheme of town and country planning should be worked out for the whole country'.[8] A Swindon man wrote more strongly, echoing both pre- and post-war concerns: 'it should no longer be possible for further encroachments by speculative builders without due regards to the needs of the people'.[9] Such comments were often coupled with essentially urban demands for access and the creation of national parks, often seen, as was indeed proposed, as a national war memorial. As a Newcastle woman wrote:

> There must of course be careful planning of both urban and rural amenities...the National Trust might be put in charge of all land, which it is desired to preserve as national monuments, it could combine with similar bodies or absorb them. The whole coastline of England...should be placed under its care together with the Lake District and Snowdonia.[10]

Many were also concerned about the decline of village life and the drift from the land. 'I would like to see', wrote one, 'the villages centres of activity and communal life, with amenities as good as those in the towns. For this a land policy is needed which does not drain all the youth away to the towns.'[11] The need for a 'land policy' was clear to many respondents. Some wanted land nationalization, but more saw smaller farms, together with some kind of policy which would open up farming to new blood, as essential, for example from Kent: 'I would like to see it possible though, for all who want it, to be able to have small holdings'.[12] Or, most simply a woman from Norfolk: 'I wish things could be arranged that my husband and I could have a farm after the war.'[13] Against this, although at the time it did not appear contradictory, was an almost universal belief in the need for continued government support and subsidy for agriculture to keep up production. In a sense the contradiction was dealt with by coupling the demand for government intervention with strict cropping controls, smallholdings and often land nationalization. In the most sophisticated of these views a new 'farming ladder' was envisaged, with large farms buying and selling co-operatively, using machine pools to produce subsidized cereals, and with national minimum wages paid. Alongside these would be smaller farms, often dairying or market gardening, acting as 'starter' units to revive village life, preventing flight from the land. A careful town and country planning authority would stop 'urban encroachment'.[14]

Interestingly, the perceptions sent to M-O in 1942 that the war was a central moment of change were frequently shared by the group answering the 1995 directive. A retired teacher from North Wales wrote: 'finally have I seen the countryside change? Heavens, yes, more so since World War II than ever I think, so much because of Government policy.'[15] A younger teacher wrote simply: 'it all began with directed production in World War II'.[16]

Despite wartime criticisms the years between 1945 and the 1980s were generally seen as a success for agriculture, and writers have begun to analyse changing views of this period (Howkins 2003; Wilson 2001). Based on the policy of subsidy enshrined in the 1947 Agriculture Act the industry was, as Howard Newby writes: 'spectacularly successful; farm productivity increased fourfold in the four decades after 1939, and by 1983 Britain had become virtually self sufficient in temperate foodstuffs' (Newby 1987: 186). In the 1950s and 1960s those criticizing this change were increasingly seen as marginal or cranky in face of consumer demand for cheap food and season-round variety. As Tracy Clunies-Ross writes in her study of post-war organic farming, 'during the 1950s and 1960s they [organic producers] became a discredited group who were considered to have nothing serious to offer' (Clunies-Ross 1990: 209).

Nor did academic work, whether economic, social or scientific, challenge these views. Economic historians such as Holderness accepted the prevailing wisdom, even if later accounts raised worries on some environmental issues (Holderness 1985; Martin 2000). No such problems were diagnosed in mainstream agricultural writings before the late 1970s. In 1979 the Seventh Report of the Royal Commission on Environmental Pollution could conclude, 'there is no doubt that the use of pesticides is essential to maintain crop yields and therefore keep down the costs of agricultural products to consumers' (1979: 6). Nor did sociology say much about large-scale agriculture, staying with studies of 'marginal' agricultural regions seen as threatened 'traditional societies'. British-based anthropological studies similarly followed in the traditions of Arensberg and Kimble's work on peasant society in Ireland (Wright 1992). For general rural issues government policy and academic study was piecemeal. The 1947 Town and Country Planning Act and 1949 National Parks and Access to the Countryside Act laid down parameters, but the extent to which, especially from the late 1950s, these were constantly under attack, particularly in the south-east, is clear from the records of the then Council for the Preservation of Rural England, held at the Rural History Centre, University of Reading (Blunden and Curry 1989). Continuing urban out-migration meant that parts of 'rural England, which had been agricultural England, swiftly became middle class England' (Newby 1987: 222). Here there was more academic interest through studies by sociologists such as Pahl and anthropologists such as Strathern, usually couched in terms of insider/

outsider conflict within the social construction of 'the village' rather than in the confrontation of bigger issues (Pahl 1965; Strathern 1981).

From the 1960s this largely positive view came to be questioned. Criticism came initially not from organic growers or green theorists but from new environmental concerns among existing organizations, especially those concerned with wildlife. A key text was Rachel Carson's American study *Silent Spring* (1962). Carson's concern as a scientist was to produce a case for the modified use of pesticides, especially DDT, which, she argued, not only destroyed harmful pests but, by destroying food for birds in particular, did widespread environmental damage. However, Carson distanced herself from any wider organic arguments insisting that 'she was not arguing against the use of all chemicals, she did not want to turn the clock back, she merely wanted more careful consideration to be given to the harm that could be caused by the indiscriminate use of ever more lethal chemicals, especially where they combined to form a cocktail effect' (Clunies-Ross 1990:168). Whatever Carson wished, the effect was to alert pressure groups in Britain, especially those connected with bird life, to dangers of chemical agriculture. From *Silent Spring* onwards, environmental awareness began to grow, not primarily in the form of an organized 'green' politics but in a generalized sense of something going wrong with agriculture and the countryside.

A good deal of such general unease is certainly based on a misguided view of the rural past, a fact not lost on the farming lobby. *Farmers' Weekly* noted in February 1994 that: 'most consumers romanticize a British countryside that never was, yet they are wholly reliant on modern farming for low-cost, high quality food' (*Farmers' Weekly* 1994: 5). Nevertheless from the late 1960s there were real concerns which could not be dismissed as simple 'golden ageism'. Most striking and visually powerful was the loss of hedgerows. Between 1946 and 1974 farmers removed a quarter of the hedgerows in England and Wales, some 120,000 miles. In some areas it was worse; in Norfolk 45 per cent of hedgerows were removed, mainly for increased barley production, and in Cambridgeshire 40 per cent (Shoard 1980: 34–41).

This is clearly reflected in the 1995 M-O directive, with 55 per cent of men and 67 per cent of women believing that the countryside had changed for the worse in their lifetimes. Of all changes described as 'bad', hedgerow removal came top, with 34 per cent of male and 26 per cent of female respondents singling it out as the most obvious change since the war. Frequently hedgerow loss was equated, largely correctly, with the introduction of large-scale machine farming. A Surrey man writes: 'undoubtedly the countryside has changed dramatically over time as more and more machinery is introduced and hedges get ripped out to make way for more intensive forms of farming'.[17] A retired librarian from a Norfolk village writes: 'My admittedly rather jaundiced view of Norfolk's countryside is reinforced by the money-grubbing vandalism which resulted [in] the grubbing up of the lovely hedgerows which sustained so much wildlife'.[18] A working-class man who grew up in the London suburbs remembers cycling in Sussex and Surrey just after the war:

> Further out in the country we used to see patchworks of small fields each being used for a different crop or grazing. Now all the hedgerows have been grubbed up and larger fields made for easier management with agricultural machines.[19]

These statements point to why the loss of hedgerows is so central to the notions of a changing countryside. First, they are believed to be homes for wildlife; again evidence overwhelmingly supports this view. The British Trust for Ornithology's annual monitoring has revealed:

> a dramatic fall in population numbers for a whole range of farmland [birds] over the past twenty-five years – the tree sparrow down 89 per cent; the bullfinch down 76 per cent; the song thrush down 73 per cent; the spotted flycatcher down 73 per cent; lapwings down 62 per cent; skylarks down 58 per cent; linnets down 52 per cent. Population numbers of the corn bunting are now too low to be routinely monitored. (Harvey 1997: 24)

Secondly, it is not only the destruction of animal and bird life by the physical destruction of habitats but, more importantly, the interpretation of that destruction. Paul Brassley has argued in an important paper that changes in the 'ephemeral landscape' – for example field boundaries such as hedges and walls, as well as cropping – are central to perceptions of the countryside. They, as much as the great subjects of space and vista which so influence artistic theory, affect how we respond to the countryside. Brassley concludes:

> the ephemeral components of the landscape have a major, and hitherto unrecognised, influence on the way in which it is perceived and valued. Moreover, many of the ephemeral components of the agricultural landscape have been subject to extensive changes over the last sixty years. Consequently, when ordinary people articulate concerns of the rapidity of agricultural change, it may be the changes to the ephemeral landscape, which are the root cause of their concerns. (Brassley 1998: 129)

This is borne out clearly in the M-O material, many noting changes in meadowland. A woman who grew up on a Dorset farm before the Second World War writes of 'the meadows in June like Swiss Meadows, full of a variety of flowers, herbs and grasses so good for the cows when made into hay'.[20] But the sense of visual change extended much further for an Essex health education officer: 'thatched barns replaced by silo's [sic] and battery chicken houses, different crops – we never used to see bright yellow fields of oil seed rape and now blue ones of flax – even the pigs have changed their shape'.[21]

Linked to problems in the 'ephemeral landscape' were the increased use of machinery, pesticides and herbicides, widely blamed for the deterioration of country life. A woman writes from the outskirts of Norwich: 'the land between our house and the bypass is farmland. Intensively farmed unfortunately. During the summer potatoes get drenched in chemicals every 7 days or so...The lapwings have long gone since all insects both good and bad have been zapped out of existence.'[22] Perceptions of landscape change here are less related to large cultural theories of, for example, national identity and rural landscape, than to 'real' changes in the countless small elements which constitute a landscape within, or perhaps alongside, the broader brush strokes of the theoretical construction. We might think of the statement made by one woman from the M-O group that 'now' all the cows were black and white whereas in her childhood there were 'different' colours. This is certainly true. The domination of the British dairy herd by the Friesian is a post-war agricultural change. But for her it marks in a profound, personal and real way, the standardization of landscape under the influence of agribusiness, concealed by continuities within the apparently permanent landscapes of, for example, the South Downs. Here, although change is real enough, it is hidden by two considerations. First, the ideology of the landscape of the south stresses continuities in the face of change (Howkins 1986). Secondly, the simple fact that the landscape's shape and contours remain the same, as long as it remains unbuilt on – but the cows are black and white, the pigs are a different shape and the hedgerows have gone.

These changes are perceived to be the result of farming practice, as were the majority of reasons given for deterioration in country life by the M-O group. The prime movers were, if only by implication, those who worked the land. Only 4 per cent blamed pressure from the government, or the EU/CAP, though a strongly worded example of this came from a woman born in Norfolk in 1932 who spent much of her life until the late 1950s overseas:

> On returning from overseas in the late 1950s I was amazed at the destruction of hedgerows, grubbing out has continued ever since, starting during WW2 to increase food production acreage govt. funded and since the late '60s CAP funded. Now 'setaside' is with us, hedges being replanted, golf courses laid out, land rescheduled for road building – all subsidised by the taxpayer...[23]

The surprisingly low figure of those 'blaming' Europe is duplicated in the work on live animal exports based on the same directive (Howkins and Merricks 2000). It is difficult to know why this should be, given the consistently anti-EU stance of much of the British media, but it could be a result of the broadly middle-class bias of the sample, although that in itself assumes that the middle class is more pro-European – by no means a foregone conclusion.

A larger group of respondents, 10 per cent of women and 16 per cent of men, showed positive hostility to farmers. These figures seem fairly low, but only 28 per cent of men and 12 per cent of women were prepared to blame anybody, with farmers easily the largest group identified. All the perceived worst problems of the countryside were associated with farming practice, and at the extreme farmers were seen as fat cats, cushioned by subsidies and destroying the countryside because of greed:

> I think the farmers have been cushioned too long by subsidies and it is about time they suffered financially as the rest of the working population have had to recently.[24]

> Sadly though, the chief change I have seen in my life is that farmers are no longer in sympathy with the land. Their activities have been dominated by politics and economics.[25]

> My views on farming changed many years ago when the prairiefication of East Anglia started, and agri-business was born. Once the cost accountants come in morality, compassion and tradition go out of the window.[26]

> It is all big business now, not a farmer making a living for him and his family in a small way... Personally I think it is only the farmers who have benefited from being in the common market, in fact I would go as far as to say that the farmers have got very rich at the expense of the rest of us. I have no sympathy with them at all.[27]

While there is more material within the M-O archive on changes in agriculture, I want to turn now to changes in 'country life' not directly attributable to farming. The most obvious changes stressed by respondents are urbanization (noted by 26 per cent of women and 17 per cent of men), and road-building (noted by 19.5 per cent of women and 19 per cent of men) as 'bad' changes. A woman living on a post-war estate built in the countryside on the outskirts of Leeds wrote that, when she moved, there were 'several patches of woodland left to grow, dotted across the estate and grassy areas too'. But since the 1960s, 'our countryside has been pushed further and further away from us, over the years the green land has been taken for private housing estates and businesses. Some of our grassy areas on the estate are now covered with a Sainsbury complex... So I'm back to feeling a "townie" again.'[28] For some, government had failed to protect the countryside, a Nottingham man writing that: 'Green belt protection, which seems to have been progressively watered-down and withdrawn by the government over the past years is essential if we are to avoid further urban development at the expense of the countryside and everyone's quality of life.'[29]

Another key area was the perceived deterioration in village life, coming mainly from people living in villages or suburbs. Complaints echo much post-1970s concern for rural deprivation. A Chelmsford man wrote: 'villages used to have shops, at least of some sort as well as frequent buses to towns. Now all this is gone.'[30] A man who had lived in Devon for a time wrote of: 'extreme poverty; most of the young people were existing at subsistence level'.[31] There was also a real sense of the 'local/incomer' battles made so much of by rural sociologists. A women born in Polegate in Sussex and still living there wrote:

> in the early sixties when people started to build new estates the village began to expand. I remember feeling a bit resentful of the people who moved in. They were mainly people who had retired from London. They were different to the local people I knew.[32]

Equally familiar is the anger from a North Lincolnshire village at incomers pushing up house prices and forcing out 'locals':

> The small hamlet that was once lived in by farm workers is now inhabited by wealthy middle class professionals...my friends parents who had retired (who had lived in a semi-tied cottage) ended up in a council flat in town.[33]

Some, however, believed that things had improved since the war, and 4.5 per cent of men and 1 per cent of women (including three farmers' wives) felt that farmers were doing a good job; although a number of others felt that farmers were pushed by forces outside their control to do 'bad'. A larger group felt that aspects of the countryside had changed for the better. Among women 4.5 per cent felt there was less rural poverty than in their childhoods, while a small number of men and women felt there was greater access to the countryside than there had been, especially before the war.

The economic history of post-war agriculture is seen in terms of success, albeit occasionally qualified. Such accounts stress increased output and labour productivity and a wider variety of crops, while persisting with arguments for agricultural stewardship of the countryside. However, even given the problems of representation with the group involved at the core of this chapter, it is clear that the experiences of those living through this transformation suggest a different view. It is widely felt that the costs of change have been too high in terms of quality of life. Thus 68 per cent of women and 57 per cent of men felt unequivocally that rural life had deteriorated in their lifetimes. Seen like this, the history of agriculture since the Second World War is far from successful, in public opinion at least.

The reaction of the agriculture industry to such late twentieth-century criticisms has been twofold. First, blame has been passed to urban demands for cheap food. Secondly, it has been argued that the urban majority do not 'understand' country life and agriculture. In this view the countryside is essentially 'England', an embattled minority surrounded by an alien and cosmopolitan society seeking to destroy its ways and 'freedoms'. Time and again the rhetoric is that of 1940, with the countryside representing Britain standing alone in the face of tyranny. Not for nothing was one of the main slogans of the 1998 Countryside March 'Resist the Urban Jackboot'.

These divisions appear fundamental to any consideration of the current British, and especially English, countryside. England, we could argue, is the first post-industrial nation as she was the first industrial one. The contradictions revealed are powerful, despite the national urbanizing and supposedly post-modernist trends since the 1980s. It is clear that a huge number of people want to live in rural areas but have no way of fulfilling that dream. Even more regard the countryside as a site of leisure. It is largely these desires which produce the kind of reactions charted in the replies to the 1995 M-O Directive. What is interesting is the extent to which replies to the 1942 Directive on the future of agriculture foreshadowed many of these concerns. Worries about the decline of village life and community, a wish to have access to land for recreation, and even concerns about the 'damage' done by some farming practices were all present in the middle of the war. What was different in 1942 was that the vast majority of those who wrote were broadly 'pro-farming' even when critical. Then the problem was perceived as being one of government, and its solution one of planning. By 1995 it was seen as being a problem of agriculture, and its solution, as far as one was offered at all, lay in farming putting its house in order and recognizing the needs and demands of an urban, or at least non-farming, population.

Against that there is clearly a strongly held belief among the minority (albeit a growing one) of the population who live in the rural areas that the urban world is using its power to crush much that they see as central not only to country life but to national identity. Added to this is the undoubted and obvious decline in aspects of the rural economy at the end of the twentieth century, which has produced a sense of bitterness and anger in many country districts, made worse by the outbreak of foot-and-mouth disease in 2001. Ultimately, there is a deep irony here. Both sides of the urban/rural divide believe in the same thing – the essential role of the countryside in English national culture – and both believe that they are its defenders. Both use the same images of timelessness and continuity, although they attach them to different things. For those who criticize current farming practice there is a strong contention that such practices have destroyed the traditional landscapes of Britain and seriously damaged wildlife. To those who defend modern agriculture it is the trade of farming itself, regardless of change, which is the continuity – only those who own or work the land can really understand it.

In spring 2002 these problems were yet again vexing government. At the Labour party conference in October 2001 Margaret Beckett, Secretary of State at the Department of the Environment, Food and Rural Affairs said:

> like the rest of the rural economy agriculture is subject to enormous pressure for change . . . the wider European public will no longer permit farming simply to carry on as before – let alone pay for . . . What society as a whole wants from agriculture is changing and probably changing irrevocably . . .

Government thinking on change was indicated in January 2002 in the policy review conducted by Sir Donald Curry (*Farming and Food* 2002), which presented agricultural crisis as part of a much wider crisis of rural areas. As the *Guardian Unlimited* put it: 'The familiar countryside environment – originally the product of farming – is damaged by years of intensive production and the continued future of the countryside is being put at risk' (www.guardian.co.uk/country, 30 Jan. 2002). In what amounts to a rejection of all British agricultural policy since 1939, subsidies are now to be removed from production and given to environmentally friendly and sustainable agricultural units and practices. Further, in recognition of the changed nature of the countryside, farmers are urged to look at non-farming uses for their land. The reaction of the National Farmers' Union was predictable, with its secretary Ben Gill calling the proposal on changing subsidies 'stupid, deceitful and immoral' and continuing 'we will fight it strongly' (*Guardian*, 30 Jan. 2002: 7) The outcome is still to be seen, but what Curry recognizes and what this chapter has proposed is that, whatever the quantitative evidence of British farming's post-war success, the qualitative evidence suggests that many among the non-farming population think very differently.

NOTES

1 Mass Observation Archive, University of Sussex (hereafter M-O) 'Spring 1995. Mass Observation. Directive' Part 2, 'The Countryside'. I am grateful to the trustees of the Mass-Observation Archive for allowing me to use material held by them.

2 M-O Diaries, M5376, schoolteacher, Burwash, Sussex, 13 Mar. 1941.

3 M-O Directives Sept. 1942, question 2C.

4 M-O Directives, Sept. 1942, 3013, w. Berkhamstead, housewife.

5 M-O Directives, Sept. 1942, 2954, m. Preston, civil servant.

6 M-O Directives, Sept. 1942, 2925, m. Oswestry, Shropshire, clerk.

7   M-O Directives, 1942, 2723, m. Brixton, London, no occupation.
8   M-O Directives, 1942, 1563, w. Chiswick, social worker.
9   M-O Directives, 1942, 3095, m. Swindon, no occupation.
10 M-O Directives, 1942, 2457, w. Newcastle upon Tyne, civil servant.
11 M-O Directives, 1942, 2865, w. Evesham, Worcs.
12 M-O Directives, 1942, 2892, w. Otford, Kent.
13 M-O Directives, 1942, 2873, w. Sheringham, Norfolk, housewife.
14 See e.g. M-O Directives, 1942, 3084, m. Horsham, Sussex, retired.
15 M-O Directives, 'The Countryside' Feb. 1995. H2506, m. Colwyn Bay, retired teacher b. 1920.
16 M-O Directives, 'The Countryside' Feb. 1995. W2322, m. Stone, Staffs., head teacher. b.1944.
17 M-O Directive 1995, A18, m. Addlestone, Surrey, unemployed, b. 1944.
18 M-O Directive 1995, D1606, m. Attleborough, Norfolk, retired librarian, b. 1924.
19 M-O Directive 1995, H1806 m. Woking, Surrey, factory worker, b. 1925.
20 M-O Directive 1995, R1452 w. Birmingham, retired teacher b. 1916.
21 M-O Directive 1995, B2170 w. Brentwood, Essex, health education officer, b. 1945.
22 M-O Directive 1995, C2053 w. Costessey, Norwich, librarian, b. 1953.
23 M-O Directive 1995, F1589, w. Audley, Staffs., nurse, b. 1932.
24 M-O Directive 1995, P1282 w. Lichfield, child-minder, b. 1938.
25 M-O Directive 1995, W2322 m. Stone, teacher, b. 1944.
26 M-O Directive 1995, R470 m. Basildon, lorry driver b. 1934.
27 M-O Directive 1995, R1468 w. Derby, factory worker, b. 1923.
28 M-O Directive 1995, T540 w. Leeds, clerk. b. 1927.
29 M-O Directive 1995, C2717 m. Nottingham, engineer b. 1966.
30 M-O Directive 1995, A883, m. Chelmsford, Clerk b. 1933.
31 M-O Directive 1995, C2750, m. Hassocks, Sussex, unemployed b. 1969.
32 M-O Directive 1995, M1498, w. Polegate, Sussex, unemployed b. 1954.
33 M-O Directive 1995, W2538 w. Cleethorpes, housewife, b. 1963.

REFERENCES

Bloome, D., D. Sheridan and B. Street 1993: *Reading Mass-Observation Writing: Theoretical and Methodological Issues in Researching the Mass-Observation Archive.* University of Sussex: Mass-Observation Archive Occasional Paper no 1.

Blunden, J., and N. Curry 1989: *A People's Charter. Forty Years of the National Parks and Access to the Countryside Act 1949.* London: HMSO.

Brassley, P. 1998: On the unrecognised significance of the ephemeral landscape. *Landscape Research*, 23(2), 119–32.

Brassley, P. 2000: Output and technical change in twentieth-century British agriculture. *Agricultural History Review*, 48(1), 60–84.

Brown, J. 1987: *Agriculture in England: A Survey of Farming 1870–1947.* Manchester: Manchester University Press.

Clunies-Ross, T. 1990: Agricultural change and the politics of organic farming. Unpublished Ph.D. thesis, University of Bath.

Conford, P. 2001: *The Origins of the Organic Movement*. Edinburgh: Floris.

Countryside Agency 2001: *The State of the Countryside 2001*. Wetherby: Countryside Agency, 6.

Dyck, C. I. 1992: *William Cobbett and Rural Popular Culture*. Cambridge: Cambridge University Press.

Farmers' Weekly 11 Feb. 1994.

Farming and Food. A Sustainable Future Jan. 2002: Report of the Policy Commission on the Future of Farming and Food, chaired by Sir Donald Curry. London.

Harvey, G. 1997: *The Killing of the Countryside*. London: Cape.

Holderness, B. A. 1985: *British Agriculture since 1945*. Manchester: Manchester University Press.

Howkins, A. 1986: The discovery of rural England. In R. Colls and P. Dodd (eds), *Englishness. Politics and Culture 1880–1920*. London: Croom Helm, 62–88.

Howkins, A. 2003: *The Death of Rural England: A Social History of the Countryside since 1900*. London: Routledge.

Howkins, A., and L. Merricks 2000: 'Dewy-eyed veal calves': live animal exports and middle-class opinion, 1980–1995. *Agricultural History Review*, 48(1), 85–103.

Martin, J. 2000: *The Development of Modern Agriculture: British Farming since 1931*. London: Macmillan.

Murray, K. A. H. 1955: *Agriculture: History of the Second World War*. United Kingdom Civil Series. London: HMSO and Longman.

Neeson, J. 1993: *Commoners: Common Right, Enclosure and Social Change in England 1700–1820*. Cambridge: Cambridge University Press.

Newby, H. 1987: *Country Life. A Social History of Rural England*. London: Weidenfeld & Nicolson.

Pahl, R. 1965: *Urbs in Rure*. London: Weidenfeld & Nicolson.

Population Trends 1987: issue 48(Summer), 41.

Royal Commission on Environmental Pollution 1979: *Seventh Report*. London. HMSO.

Shoard, M. 1980: *The Theft of the Countryside*. London: Temple Smith.

Stanley, N. S. 1981: 'The extra dimension': a study and assessment of the methods employed by Mass-Observation in its first period, 1937–40. Unpublished Ph.D. thesis, Birmingham Polytechnic.

Strathern, M. 1981: *Kinship at the Core. An Anthropology of Elmdon, a Village in North-West Essex in the Nineteen-Sixties*. Cambridge: Cambridge University Press.

Williams, R. 1973: *The Country and the City*. London: Chatto &Windus.

Wilson, G. A. 2001: From productivism to post-productivism . . . and back again? Exploring the (un)changed natural and mental landscapes of European agriculture. *Transactions Institute of British Geographers*, 26(1), 77–102.

Wright, S. 1992: Image and analysis; new directions in community studies. In B. Short (ed.), *The English Rural Community: Image and Analysis*. Cambridge: Cambridge University Press, 195–217.

Part II

Sites of British Modernity

Chapter Six

'A Power for Good or Evil': Geographies of the M1 in Late Fifties Britain

Peter Merriman

Movements, spaces, commodities and technologies associated with the production and consumption of motor vehicles had a notable presence in debates about modernism, modernity and modernization in the West during the course of the twentieth century (Kern 1983; Ross 1995; Sachs 1992; Thrift 1996). In this chapter I focus on a series of specific spaces associated with car travel in Britain, tracing out the geographies of the construction, design and use of the first sections of Britain's M1 motorway in the late fifties, a time when car ownership was increasing at a dramatic rate and being promoted and celebrated as a pleasure, freedom and necessity in an increasingly affluent society.[1]

The M1 formed a key element in the post-war roads programme announced by Labour's Minister of Transport, Alfred Barnes, on 6 May 1946 (*Parliamentary Debates* 1946), but while its construction can be located within a discourse of reconstruction, private syndicates and professional institutions had been proposing schemes for a London to Birmingham motorway since 1902 (see Merriman 2001). During the late 1920s and the 1930s many British engineers, planners, politicians, landscape architects and preservationists praised Italy, Germany and the USA for their well-planned, safe, fast, modern motor roads (Matless 1998). Despite concerns about the politics and different senses of modern order which were bound up with the German *Autobahnen*, a delegation of 224 politicians, engineers and planners organized by the Automobile Association, Royal Automobile Club and British Road Federation toured Germany's roads for 10 days in September 1937 (*The Times* 1937; Merriman 2001; see also Dimendberg 1995; Rollins 1995; Shand 1984). After the Second World War British engineers studied the design of motorways across northern Europe, but despite ongoing pressure from lobby groups such as the British Road Federation for a start to be made to the roads programme,

Britain's continuing economic downturn, marked by the sterling crisis of 1947 and cutbacks in capital investment programmes throughout the early fifties, meant that plans to reconstruct the nation's roads were postponed. Government efforts were aimed at alleviating the housing shortage, developing the National Health Service, building new schools, sustaining the armed forces and nuclear weapons programme, and reducing national debt by limiting domestic consumption and increasing exports in manufactured goods (Gardiner 1999; Marwick 1996). The aesthetic styles, materials and experiences associated with modern buildings, technologies, practices and mass-produced consumer goods appeared in sharp contrast to the austere realities of many people's everyday lives, but the latter half of the fifties saw increasing affluence for many as taxes fell, average incomes rose, and near-full employment was achieved. The Conservative prime minister Harold Macmillan was confident enough to state in 1957 that 'most of our people have never had it so good': 'Go round the country, go to the industrial towns, go to the farms and you will see a state of prosperity such as we have never had in my life-time – nor indeed ever in the history of this country' (quoted in Sampson 1967: 158).

Commercial television was launched in 1955, the Consumers' Association formed in 1957, and hire purchase restrictions were lifted in 1958. The public were buying cars, refrigerators, televisions and telephones in ever greater numbers, and these were becoming cheaper in real terms (Marwick 1996). It was in the midst of these changes in late fifties Britain that the design, construction and experiences associated with the M1 were seen by many to be novel and distinctly modern. At a time when modernist principles were being utilized in the planning, design and landscaping of large-scale projects such as new towns, housing estates, schools and power stations, the public and preservationists showed a considerable degree of faith in experts who promised to construct motorways that would fit into the English landscape – although as the design and construction of the M1 would illustrate, it was not uncommon for designers and commentators to disagree about what counted as good modern design. The purpose of this chapter is to explore the modernism and modernity of this motorway, where modern spaces, aesthetics and experiences can be seen to emerge through the folding of different times, spaces, movements and subjects into topological formations that are hybrid, contingent, contested and partial rather than monolithic, fixed or epochal (cf. Conekin et al. 1999; Ogborn 1998; Thrift 1996).

Designing

The consulting engineers Sir Owen Williams & Partners started conducting preliminary surveys of potential routes for the M1 in 1951, although government cutbacks meant that the decision to proceed with detailed designs was delayed until February 1955 (*Parliamentary Debates* 1955). Sir Owen had been a pioneer of concrete reinforcement techniques in Britain in the 1910s and 1920s (Newby 1986), and was widely praised by British and European modernist architects and critics for a series of major public buildings in the 1930s – notably the Daily Express buildings, the Pioneer Health Centre in Peckham, and the Boots Wets factory in Beeston near Nottingham (Stamp 1986; Yeomans 2000). In *An Introduction to Modern Architecture* (1940) J. M. Richards, acting editor of *The Architectural Review*, celebrated Sir Owen's Boots factory as one of the best and earliest examples of large-scale modern architecture in Britain, while the architectural and cultural critic Reyner Banham admired his pre-war structures for their 'pure construction, unsullied by aesthetic intentions' (1960: 784). Banham was not so complementary about Sir Owen's post-war work. The bridges on the M1 were seen to demonstrate a loss of purity and functionality, and the adoption of an 'anti-aesthetic' approach (Banham 1960: 784) (see figure 6.1). These were 'the ugliest set of standardised structures in Britain . . . coarse, cheap bridges' (Banham 1972: 242). One of the few critics who spoke favourably of them was the President of the Royal Institute of British Architects, Basil Spence, who explained how 'their breadth and strength' reminded him 'of some of the great Roman works' (1959: 36). But other critics, including John Betjeman, Ian Nairn, Nikolas Pevsner and the landscape architects Sylvia Crowe, Geoffrey Jellicoe and Brenda Colvin, criticized the bridges for their heaviness, bulk, excessive standardization, ornamental parapets, dangerous central supports, and interruption of the flow of the landscape both along and across the road (Merriman 2001). Sir Owen Williams had himself stated that the bridges

Figure 6.1. Two-span overbridge across the M1 motorway designed by Sir Owen Williams & Partners
Source: Photograph reproduced courtesy of John Laing Construction.

were simple, modern, functional designs of a 'bold, massive manner' (quoted in John Laing & Son Ltd. 1958: 5), but this was not a modernism and functionalism admired by his critics. At a time when pre-stressing and pre-casting techniques were being employed to construct unobtrusive, light and clean-lined bridges, Sir Owen's decision to utilize *in situ* cast reinforced concrete was deemed rather outdated, slow and costly. It resulted in what many critics felt to be an inappropriate modernist aesthetic; 'MODERN ARCHITECTURE only with reservations' was Pevsner's verdict in his guide to the buildings of Northamptonshire (1961: 66).

While the modernism of Sir Owen's bridges was widely criticized, the design of other aspects of the motorway – including the signposts, service areas and roadside planting – also emerged as points of contention. In the midst of what were often quite complex debates and disagreements was the government's Advisory Committee on the Landscape Treatment of Trunk Roads. The committee was formed by the Minister of Transport and Civil Aviation in April 1956, and despite a background of disagreements between its constituent organizations, the members were unified in their criticisms of the planting plans prepared for the M1 by Sir Owen Williams and Partners.[2] They expressed concern about the overall landscape design of the motorway and the appearance of the bridges, but civil servants ensured that it was beyond the remit of the committee to prevent any delays in construction over these matters:

> it has taken . . . several years to find a satisfactory line. If a road of this descrip-tion were then to be submitted to a Committee who would no doubt want to go over it from end to end and criticize it from an aesthetic point of view and the consultants had then to examine their alternative suggestions, I am afraid that the making of the scheme and the actual construction would have to be very considerably postponed. (Haynes, Minute 15 23/12/1954, Public Record Office (PRO) file MT 121/576)

The committee's powers were limited to more superficial aspects of the design, and it rejected the planting schedules prepared by Sir Owen's consultant on forestry and landscape, A. P. Long, and his assistant, A. J. M. Clay. The plans had included ornamental shrubs and flowering plants such as fuchsia and forsythia, which were deemed to be of an urban or 'semi-urban character' and 'misplaced in real countryside' (Minutes LT/M15, 17/7/1957, PRO MT 123/59). In contrast, the committee felt that the motorway should be treated as part of this 'real countryside', with indigen-ous species, common varieties and simple layouts planted for functional rather than decorative reasons. Functional planting could soften the lines of embankments, cuttings and bridges, screen unsightly views from the road, reduce 'parallelism', monotony and dazzle, and even improve the views of the motorway from the surrounding countryside. Expert commentators

believed that this modern motorway could fit easily into and even enhance
the English landscape, providing drivers with new experiences of the countryside – what Brenda Colvin described in *The Geographical Magazine* as 'a
new look at the English landscape' (1959: 239). The simple, functional
design of the motorway would be free from distractions and reflect these
new experiences, rhythms and appearances. As the Landscape Advisory
Committee member Sir Eric Savill stated in February 1958, 'a fast motorway is not a place for the encouragement of interest in flowering shrubs.
"Eyes on the road" should be the motto!' (Letter 6/2/1958, PRO MT 121/
78). The design of the motorway should be orderly and attractive but not
distracting, while modern, functional architecture and engineering must be
accompanied by functional arrangements of indigenous plant species of a
colour (essentially green), texture, size and shape that would knit together
the motorway, moving vehicles *and* the surrounding countryside:

> Just as the functional engineering of the road can produce its own landscape,
> indicative of speed, so can the functional use of planting produce the link
> between the landscape of speed and the landscape of nature. (Crowe 1960: 95)

The Landscape Advisory Committee made similar recommendations for
the planting of the first two service areas at Newport Pagnell in Buckinghamshire and Watford Gap in Northamptonshire. While the activities of
travellers and speed of movement were different from the motorway proper,
indigenous species were still recommended for functional and aesthetic
reasons. Committee members were concerned that ornamental species
might excite rather than relax drivers who were resting after a long journey,
while objections were made to plants whose colours 'may clash with that of
the petrol pumps' (Report LT/167, PRO MT 121/360). Designs for the
main buildings at both of the service areas were prepared by architects
appointed by the developers, but the Landscape Advisory Committee
expressed concern at the appropriateness of certain modern shapes in the
English landscape following a report by the Council for the Preservation of
Rural England's representative George Langley-Taylor:

> I find it difficult to comment [on the design] because I fear that my objection
> to the long flat roof may be interpreted as an objection to modern architec-
> ture. Frankly I do not like it because I feel however right it might be as a
> modern building this long straight line is bound to be a jar on the landscape
> and I feel most strongly that in dealing with our motorways we should try to
> achieve a sympathy with the landscape and avoid introducing any 'shock' in
> our designs. (Letter 26/1/1960, PRO MT 121/182)

Langley-Taylor's comments related to the police post proposed for Watford
Gap service area, but the Landscape Advisory Committee was critical of

the designs for all of the service area buildings and the lack of co-ordination between the different architects designing the police posts, main buildings, and maintenance depots. While Sir Owen Williams & Partners designed the maintenance depots and service area footbridges, the architects of the main buildings were appointed by the developers, and the police posts were designed by the county architects. Both the Royal Fine Art Commission and Landscape Advisory Committee complained to civil servants about the piecemeal way in which designs were submitted to them (see PRO MT 121/182; MT 121/359), and further criticisms emerged in the architectural press when the two service areas were completed in the autumn of 1960. *The Architects' Journal* described Watford Gap as 'a commonplace design which does not auger well for future motorways' (Astragal 1960: 417), while Newport Pagnell service area was criticized by Raymond Spurrier for its 'nondescript buildings and irresolute planning' which brought the 'usual subtopian results' (1960: 406). The solution to such visual confusion must be 'the guidance of a master eye trained in the art of arranging and organizing shape and colour and pattern' (1960: 408). The Landscape Advisory Committee and Royal Fine Art Commission recommended the appointment of a co-ordinating architect, and civil servants received names of potential candidates. However, the Ministry decided that the service area buildings would not 'be of great importance' or 'worthy of the attention of architects of the eminence of those suggested by the President of the R.I.B.A.' (Mills, Letter 4/4/1960, PRO MT 121/359).[3]

As new motorways and service areas were opened during the 1960s and 1970s, the reputation of the M1 further declined. The architectural commentator Ian Nairn, writing in 1975, claimed that motorways had emerged as 'one of the few genuinely collective and genuinely hopeful parts of design in Britain', counteracting the unplanned, chaotic, subtopian sprawl he had documented in a special 'Outrage' issue of *The Architectural Review* in 1955 (1975: 329). However, Nairn felt that this had not been achieved on the earliest stretches of the M1. Its inappropriate and unco-ordinated plans and designs appeared in sharp contrast to other stretches of motorway; Reyner Banham described it as 'the ugliest piece of motor road in the world' (1972: 242).

Constructing

While architectural and design commentators criticized the planning and design of the M1, many other individuals celebrated the novelty, spectacle and modernity of Britain's first major motorway. The contracting engineers John Laing & Son Ltd promoted the motorway as a space of modern construction methods, organization, and work practices, and as the biggest civil engineering contract carried out in Britain. The engineering historian, railway enthusiast and 'English organicist' Tom Rolt authored Laing's commemorative booklet *The London–Birmingham Motorway*, where he celebrated 'the greatest concentration of mechanical power ever mustered on one contract in this country' (1959: 6; for discussion of Rolt's organicist perspective on the English landscape, see Matless 1998: 149–53). Rolt had just finished his research for a biography of George and Robert Stephenson, and he found it instructive to draw comparisons between the methods and movements of the 4,700 men and 182 modern machines used to construct the motorway and the 25,000 men with only 'primitive equipment' who built Robert Stephenson's London to Birmingham railway in the 1830s (1959: 7). This history of civil engineering enabled Rolt to highlight the modernity and technical sophistication of the construction process, but the 'famous and familiar' names who manufactured the machines discussed by Rolt chose to celebrate the construction of the motorway through more futuristic associations (1959: 25). Caterpillar and Blackwood Hodge both placed advertisements in *The Times* on the opening day of the motorway, 2 November 1959. Blackwood Hodge suggested that their Euclid earth-moving machines had laid 'the foundations of tomorrow's brave new world' in the midst of 'the rush and roar of the twentieth century', constructing this futuristic 'highway to tomorrow' so as to link the nation's 'factories, ports and cities' (1959: 8). This was, as Caterpillar stated in its advertisement for the D8 tractor (see figure 6.2), 'the beginning of a new era': an era of 'progress and opportunity' where modern motorways would both reflect and help reinforce 'the Nation's prosperity' (1959: 5). Caterpillar was 'helping to build a better Britain', and this North American company was now building its D8 tractor in Glasgow, contributing to the nation's economic growth and the progress and reconstruction of a modern society.

The motorway was folded into a complex narrative of Britishness, emerging as a symbol of the greatness of 'the nation' and the British engineering expertise which helped construct it. As J. M. Laing stated during one of the Minister of Transport and Civil Aviation's visits to the construction site in June 1959, the 'British civil engineering operative is the best of his class in the world' (quoted in John Laing & Son Ltd 1959: 8). However, this story of Britishness begins to unravel when the biographies and nationalities of the workforce are explored. The multinational workforce employed by the contractors included Canadians, South Africans, Indians, Jamaicans, Poles and Hungarians, as well as large numbers of English and Irish labourers, and smaller numbers of workers from Scotland and Wales. Histories of migration, displacement, empire, war, and post-war employment and immigration policies complicate references to the 'Britishness' of construction, expertise, and of the motorway

LONDON—BIRMINGHAM MOTORWAY:
Complete and In Operation Nov. 2nd, 1959

Today is the beginning of a new era. But only the beginning. Our highways will reflect the Nation's prosperity, save the Nation more than they cost to build, benefit each and every one of us—and we need them NOW.

Caterpillar has contributed much towards these new avenues of progress and opportunity—and will contribute more. For Caterpillar earth-moving equipment is as reliable and sure as progress itself.

The Caterpillar D8 Tractor, now built in Great Britain, is the undisputed leader in its class. Manufactured to strict quality control standards the D8 moves earth cheaper and faster than ever before.

CATERPILLAR

Helping to build a Better Britain

Figure 6.2.   Advertisement for Caterpillar, 1959
Source: Reproduced by kind permission of Caterpillar Inc.

itself. The multinational and multiracial character of the workforce was highlighted and celebrated in the public relations films and booklets produced by John Laing & Son Ltd, although Rolt described the disappearance of certain differences as construction progressed: 'The men of many races who made up the Motorway team became almost indistinguishable as the sun burnt English skins to the colour of the Jamaicans or the turbaned Sikhs who worked beside them' (Rolt 1959: 49).

These seemingly harmonious relations between workers of different nationalities and ethnicities were seen to be progressive and positive at a time when the colour bar was widespread and race riots had occurred the previous year in Notting Hill and Nottingham. However, Rolt's statement of *presence* acts to create a spectacle of foreignness and non-whiteness, equating Englishness with whiteness and suggesting that white/English labourers became coloured through their association and work with black 'others' under a normalizing sun, echoing nineteenth-century debates about the 'colouring' of Jews, the working class and Irish as 'white negroes' (McClintock 1995: 52). And, as the account of one Jamaican worker revealed, discrimination and racism could lie behind these accounts of equality and harmony. The worker in question was discussing his experiences with an Irish workmate in a pub, where their conversation was recorded by Ewan MacColl and Charles Parker for the BBC Radio ballad 'Song of a Road'. The man described how he had driven bulldozers on construction sites in Jamaica and alleged that his less privileged position on the motorway was due to discrimination by his employers: 'I write in many times for a job, but just because I've been there, and they see that I'm a different nationality, they don't give it to me' (Birmingham City Archives CPA/LC86).

While John Laing & Son Ltd celebrated the working of the motorway by modern machines, promoting it as an inclusive, modern, British engineering project, the company was not simply fixed in place through a range of sites and materials – whether finished projects, construction sites and offices, charts, maps and company records circulated between and archived at specific locations, or accounts of projects and the company in film, newsletters, press releases and booklets. Rather, the company and motorway were 'materially heterogeneous', and contingently and continually assembled and performed through movements, 'talk, bodies, texts, machines, [and] architectures' (Law 1994: 2; Philo and Parr 2000). The ongoing ordering and achievement of the motorway and the company entailed a range of practices, techniques and technologies for governing the movements and desires of workers and their families; visualizing information and ideas; categorizing men, women, machines and materials; narrating the construction project and the company; transporting ideas, workers and tools; and calculating changes in such registers as profits or tarmac constituency. These practices and technologies enabled engineers to order the spaces, materials and subjects of the motorway in distinct yet partial ways, while labourers and directors would continually resist, reappropriate or rework specific techniques and technologies, performing and narrating their work in regularized yet personalized ways. One character whose movements and conduct attracted the attention of local authorities, residents and journalists alike was that of the 'lorry-driver', who travelled between the construction site and local quarries. The movements and conduct of driver and vehicle were seen to become inseparable, as local residents and Buckinghamshire and Bedfordshire councillors complained about lorry-drivers who damaged ditches and verges, ignored stop signs, broke speed limits, and caused accidents (*Chronicle and Echo* 1958).

Specific techniques and technologies for ordering the motorway enabled engineers and commentators to comprehend and manage this 55-mile-long construction site as a single project. L. T. C. Rolt's *The London–Birmingham Motorway* and Laing's films *Major Road Ahead* and *Motorway* presented narratives of construction to the public, employees and potential clients. Maps, charts and graphs provided simplified statistical overviews of the project, and Rolt was almost overwhelmed by the diverse array of drawings in the planning room at Laing's Newport Pagnell headquarters, where 'the large map . . . made progress visible at a glance' (1959: 27). One of the most popular overviews, however, was undertaken from the air as engineers, VIPs and journalists all took to the sky. Aerial photographs of the motorway appeared in national and local newspapers in 1958 and 1959, while Laing's hired BEA helicopter enabled engineers to check the movements and activities of workers and machines, and transport guests, mail and equipment along the route. The helicopter left Rolt with 'an impression . . . which no statistics, no charts, no maps . . . could possibly convey', as the near-complete structures and shapes appeared to bring order and a 'disciplined geometry' to the landscape (1959: 29–30). Modern planning, architecture and engineering combined to shape a modern English landscape (cf. Matless 1998), while the helicopter emerged as a modern technology for assembling and governing working subjects, transporting a range of materials, and visualizing the landscapes of the motorway.

Driving

> This motorway starts a new era in road travel. It is in keeping with the bold, exciting and scientific age in which we live. It is a powerful weapon to add to our transport system. But like all powerful instruments it can be a power for good or evil. (Marples 1959)

In his speech at the opening of the M1 on 2 November 1959, the new Minister of Transport Ernest Marples associated Britain's newest motorway with other novel 'weapons' or 'instruments' of a modern 'scientific age'. These oblique references to the ongoing development of nuclear or space technologies were followed with a simple warning, that drivers must 'use discipline, common sense and obey the rules' in order to prevent 'disaster and tragedy' descending on them (Marples 1959). Motorway regulations and a new Motorway Code had been developed in time for the opening of the Preston bypass motorway in December 1958, and these acted to construct the motorway driver as a somewhat hybrid and hetero-geneous subject, whose actions, mobilities and senses of being and dwelling were seen to be inseparable from specific vehicles (Urry 2000; Michael 2000). This 'vehicle-driver' had to be fast, powerful, of an accepted size and weight, centred on an inanimate source of power, have a qualified human operator, and be educated, policed, governed and serviced through a range of new techniques and technologies (*Parliamentary Debates* 1958). Britain's motorway regulations did and still do list materially heterogeneous 'mobile-subjects' that are prohibited from using motorways, including cyclists, pedestrians, agricultural vehicles, mopeds, animals, unauthorized oversized loads, learner drivers, and invalid carriages. The Motorway Code was incorporated into the Highway Code in 1959 (Ministry of Transport 1959). The Automobile Association, Royal Automobile Club, specialist magazines such as *The Autocar* and *The Motor*, and national newspapers also provided advice and motoring supplements for drivers on the M1. The authors of these documents emphasized the importance of lane discipline, vehicle maintenance, using indicators, and adhering to the motorway regu-lations, but in spite of this advice Marples expressed horror at the conduct of the first drivers on the motorway: 'I was frightened when I saw the first drivers using the road. I have never seen anybody going so fast and ignoring the rules and regulations' (quoted in Mennem 1959: 5).

Motoring journalists felt that Marples was rather hasty in his remarks, not least because they were among those testing this speed-limitless road, and one reporter was a passenger in a 3.4-litre Jaguar that passed the minister at 120 m.p.h. (*Herts Advertiser* 1959). These journalists believed that speed was not the major problem; it was drivers who lacked the skills, experience and vehicles for high-speed motorway driving who constituted the main hazard. In a satirical article in *Punch*, H. F. Ellis stated that it was the lorry-drivers 'released from the constraints of A5', the 'old fool in a worn-out soap box', and the 'normally rational people in unbalanced saloons' who were expected to move beyond their physical, mental, and technical abilities (Ellis 1959: 363). In contrast to these ill-equipped vehicle-drivers was the figure who possessed the expertise, capabilities and vehicle necessary to advise motorists on the advantages and dangers of this new motorway: the youthful, masculine British Grand Prix racing driver.

On 8 November 1959 *The Observer* published an article by Ferrari's 27-year-old Grand Prix racing driver Tony Brooks, in which he commented on the design and performance of Britain's newest motorway (Brooks 1959; Owen 2000). Brooks remarked on the behaviour of drivers and vehicles he encountered, and the 'effortless cruising of the Aston Martin', while the overall experience of driving along the motorway struck him as being decidedly different and modern:

> To drive up M1 is to feel as if the England of one's childhood ... is no more. This broad six-lane through-way, divorced from the countryside, divorced from towns and villages, kills the image of a tight little island full of hamlets and lanes and pubs. More than anything – more than Espresso bars, jeans, rock 'n' roll, the smell of French cigarettes on the underground, white lipstick – it is of the twentieth century. For all that, it is very welcome. (Brooks 1959: 5)

The motorway appeared as a distinctively modern and dis-integrating force in the English countryside. This was a scene far removed from the Landscape Advisory Committee's vision of a modern road that could be easily integrated into its surroundings. The motorway was associated here with youth and foreign tastes and styles; with the American influences and spaces criticized by Richard Hoggart in *The Uses of Literacy* (1957) being situated alongside the fashionable, Italianized practices, styles and spaces – Espresso bars, scooters and sharp tailored suits – that were emerging in late fifties London (Gardiner 1999).

While the service areas had not opened when Brooks made his assessment of the motorway, these formed key sites in the experience and performance of the motorway as a modern space and place, and Newport Pagnell service area, which was opened in 1960 and jointly owned by the Soho café, milk bar and coffee house company Forte and Blue Star garages, became an exciting 'place of pilgrimage for teenagers hoping for instant glamour':

> For young people, the new road was a concrete escape to a new kind of excitement. . . . Mr Forte's snack-bar on the M1 . . . this cosy man-made island called out to Britain's youth, the generation of teenagers who did not know there was anything special about being young but forsook the coffee bars of Soho to spend Saturday night 'doing a ton' on this long straight road. (Greaves 1985: 8)

The romantic and nostalgic reminiscences of Suzanne Greaves highlight the novelty and excitement which affluent teenagers and twenty-somethings associated with the motorway and its service areas; just a year after these novel consumption practices had been mapped out in Colin MacInnes's novel *Absolute Beginners* (1959) and the market researcher Mark Abrams's essay *The Teenage Consumer* (1959). However, this 'bright, modern and comfortable' service area, designed 'to blend with the exciting conception of the motorway itself', was intended to cater for broad range of consumers, who could eat 'popular cooked meals and snacks' in 'quick-snack' cafeterias or opt for a more leisurely meal served by waitresses in 'the Grill and Griddle restaurant' (Hartwell 1959: 13). While the service areas were designed and experienced as distinctly modern spaces to which young and old flocked, the practices and experiences of consumption associated with these spaces became bound up with the excitement surrounding the motorway as a whole. Drivers queued on the opening day and made detours to drive along this modern road; spending Sunday afternoons cruising along the carriageways in their family saloons, gazing down from bridges, or travelling out from the capital on one of London Transport's special bus trips (*The Times* 1959a, 1959b). There appears to be no indication of the popularity and longevity of London Transport's motorway trips, but the excitement surrounding the M1 did begin to wane as further sections of motorway opened across the country, and deficiencies in design standards started to emerge. As the Ministry of Transport admitted in a press release in 1962, while the 'M1 will remain the cheapest motorway that this country has built or is likely to build in the future', some of the 'compromises . . . accepted in the interests of economy both in cost and the use of land . . . went too far' (*Roads and Road Construction* 1962: 288).

Placing the Modern Motorway

Different times, spaces, materials and mobilities have been continually folded into the socio-material forms of the M1, which were performed through the movements, work, texts and memories of drivers, labourers, engineers, landscape architects, politicians, film-makers, AA patrolmen, and service area managers, as well as its more material forms 'on the ground'. The motorway was a hybrid entity or space which reflected and refracted a range of debates about the role of, and attitudes towards, the modern in late fifties Britain (cf. Conekin et al. 1999). With the consolidation of a series of professions and domains of expertise in the Second World War and early post-war Britain, civil servants and politicians provided increasing support for proposals by planners, engineers, landscape architects and other experts to build safe, fast, efficient, orderly and enjoyable British motorways. These were intended to bind together the city, countryside, regions and nation, providing order in the landscape and enabling the orderly movement of motorized citizens and goods to their destinations (Matless 1998). But, while experts espoused visions of a future reconstructed Britain, their plans for future British landscapes also incorporated traditions associated with past landscapes and foreign countries. Different traditions of engineering, design and landscape architecture were central to the social and material construction of the M1, while assertions of its Britishness tended to efface the complex geographies of the motorway. Trinidad Lake asphalt, Italian coffee, the German *Autobahnen*, and North American construction equipment were folded into the spaces of this modern British motorway, as was the physical work of labourers from Ireland, Poland, the West Indies and other countries. The experiences and actions of these labourers, along with those of engineers, designers, drivers, and other casual workers, were incorporated into narratives of the motorway in different ways. Tom Rolt's *The London–Birmingham Motorway* described the construction of a modern motorway that was forged through teamwork, expertise, modern technologies and efficient organization; architectural commentators suggested that the modernist aesthetic employed on the motorway was inappropriate and poorly conceived by Sir Owen Williams & Partners, and that the shapes of the service areas appeared out of place in the countryside; while drivers using the motorway were incorporated into narratives which emphasized the modernity of the driving experience, the novelty of consumption practices associated with these spaces, and the need for experts to instruct drivers.

The production and consumption of the motorway was associated with dynamic and mobile senses of space and place (Thrift 1999). New spatialities arose in relation to new movements, technologies and ways of driving, and drivers experienced familiar and unfamiliar stretches of motorway in diverse and multiple ways. Historical studies, as well as ethnographic work, can help to explore the diverse and often complex ways in which spaces of travel and transport are constructed as *places*, providing critiques of Marc Augé's ethnological observations on the proliferation of 'non-places' such as motorways, supermarkets and hotels where we encounter the excesses of a contemporary 'supermodern' world. (1995: 29). Further study could expose the deficiencies and generalizations of such theoretical writings – which overlook the diverse social relations, experiences and networks of actors constituted through these spaces and places – while also outlining the specific placings through which experiences of dislocation, boredom, ubiquity, solitariness, and placelessness *may* and *do* emerge.

NOTES

1   The motorway was referred to as the London–Yorkshire Motorway or London–Birmingham Motorway until July 1959, when the current numbering system was adopted and the 72 miles of motorway opened on 2 November 1959 were numbered M1, M10 and M45 (Minute 3/7/1959, Public Record Office (PRO) MT 120/64). Of the 72 miles, 55 miles were constructed by John Laing & Son Ltd.

2   The Advisory Committee on the Landscape Treatment of Trunk Roads was chaired by Sir David Bowes-Lyon (President of the Royal Horticultural Society) and included a number of independent members as well as representatives from the Royal Forestry Society for England and Wales, Institute of Landscape Architects (ILA), Council for the Preservation of Rural Wales, Roads Beautifying Association (RBA), Council for the Preservation of Rural England (CPRE), and Standing Joint Committee of the Automobile Association, Royal Automobile Club and Royal Scottish Automobile Club. A number of senior figures in the RBA (notably Wilfred Fox) had had a rather stormy relationship with the CPRE and ILA since the early 1930s (see Merriman 2001). The committee is referred to here by its informal title, the Landscape Advisory Committee.

3   The President of the Royal Institute of British Architects had suggested that Lionel Brett, Sir Hugh Casson, J. W. M. Dudding, Frederick Gibberd, Geoffrey Jellicoe, Sir Leslie Martin, Peter Shepheard and Ralph Tubbs would all make suitable candidates (Godfrey Samuel, Letter 24/3/1960, PRO MT 121/359).

REFERENCES

Abrams, M. 1959: *The Teenage Consumer*. London: The London Press Exchange.

Astragal 1960: Pull-up for socks? *The Architects' Journal*, 132, 417.

Augé, M. 1995: *Non-Places: Introduction to an Anthropology of Supermodernity*. London: Verso.

Banham, R. 1960: The road to ubiquopolis. *New Statesman*, 59, 784–6.

Banham, R. 1972: New way north. *New Society*, 20, 4 May, 241–3.

Blackwood Hodge 1959: Advertisement. *The Times*, 2 Nov., 8.

Brooks, T. 1959: The hazards of M1. *The Observer*, 8 Nov., 5.

Caterpillar 1959: Advertisement. *The Times*, 2 Nov., 5.

Chronicle and Echo (Northampton) 1958: Lorries 'speed' in narrow lanes. 18 Sept.

Colvin, B. 1959: The London–Birmingham motorway: a new look at the English landscape. *The Geographical Magazine*, 32, 239–46.

Conekin, B., F. Mort and C. Waters 1999: Introduction. In B. Conekin, F. Mort and C. Waters (eds), *Moments of Modernity: Reconstructing Britain 1945–1964*. London: Rivers Oram Press, 1–21.

Crowe, S. 1960: *The Landscape of Roads*. London: The Architectural Press.

Dimendberg, E. 1995: The will to motorization: cinema, highways, and modernity. *October*, 73, 91–137.

Ellis, H. F. 1959: M1 for murder. *Punch*, 28 Oct., 362–3.

Gardiner, J. 1999: *From the Bomb to the Beatles*. London: Collins & Brown.

Greaves, S. 1985: Motorway nights with the stars. *The Times*, 14 Aug., 8.

Hartwell, E. 1959: Provision of catering facilities. *The Guardian*, 2 Nov., 13.

Herts Advertiser & St. Albans Times 1959: No time to wave back! 6 Nov., 3.

Hoggart, R. 1958 [1957]: *The Uses of Literacy: Aspects of Working Class Life With Special Reference to Publications and Entertainments*. Harmondsworth: Pelican.

John Laing & Son Ltd 1958: Minister visits the motorway. *Team Spirit: The Monthly News Sheet Issued by John Laing and Son Limited*, 141, 5.

John Laing & Son Ltd 1959: Minister visits the motorway. *Team Spirit: The Monthly News Sheet Issued by John Laing and Son Limited,* 153–4, 8.

Kern, S. 1983: *The Culture of Time and Space 1880–1918.* Cambridge, MA: Harvard University Press.

Law, J. 1994: *Organizing Modernity.* Oxford: Blackwell.

McClintock, A. 1995: *Imperial Leather: Race, Gender and Sexuality in the Colonial Contest.* London: Routledge.

MacInnes, C. 1964 [1959]: *Absolute Beginners.* Harmondsworth: Penguin.

Marples, E. 1959: Speech of the Rt. Hon. Ernest Marples, Minister of Transport at the opening of the London–Birmingham Motorway on Monday, November 2nd, at 9.30 a.m. Official press release, copy in 'Motorways' file, Automobile Association archives, Basingstoke.

Marwick, A. 1996: *British Society Since 1945.* Harmondsworth: Penguin.

Matless, D. 1998: *Landscape and Englishness.* London: Reaktion.

Mennem, P. 1959: Motorway 1 opens – and Mr Marples says: 'I was appalled'. *Daily Mirror,* 3 Nov., 5.

Merriman, P. 2001: M1: A cultural geography of an English motorway, 1946–1965. Unpublished Ph.D. thesis, University of Nottingham.

Michael, M. 2000: *Reconnecting Culture, Technology and Nature: From Society to Heterogeneity.* London: Routledge.

Ministry of Transport and Civil Aviation and the Central Office of Information 1959: *The Highway Code Including Motorway Rules.* London: HMSO.

Nairn, I. 1955: Outrage. *The Architectural Review* (special number), 117, 363–460.

Nairn, I. 1975: Outrage twenty years after. *The Architectural Review,* 158, 328–37.

Newby, F. 1986: Williams and reinforced concrete. In G. Stamp (ed.), *Sir Owen Williams 1890–1969.* London: The Architectural Association, 13–15.

Ogborn, M. 1998: *Spaces of Modernity: London's Geographies 1680–1780.* London: Guilford Press.

Owen, O. 2000: Driving without a safety net. *The Observer,* 5 Mar., 20.

Parliamentary Debates, House of Commons 1946: Highway Development (Government Programme). 422, 6 May, cols. 590–5.

Parliamentary Debates, House of Commons 1955: Expanded Road Programme (Government's Proposals). 536, 2 Feb., cols. 1096–1109.

Parliamentary Debates, House of Commons 1958: Motorways (Traffic Regulations). 592, 23 July, oral answers, cols. 402–5.

Pevsner, N. 1961: *Northamptonshire.* Harmondsworth: Penguin.

Philo, C., and H. Parr 2000: Institutional geographies: introductory remarks. *Geoforum,* 31, 513–21.

Richards, J. M. 1940: *An Introduction to Modern Architecture.* Harmondsworth: Penguin.

Roads and Road Construction 1962: The M1 motorway. 40, 288.

Rollins, W. H. 1995: Whose landscape? Technology, Fascism, and environmentalism on the National Socialist Autobahn. *Annals of the Association of American Geographers,* 85, 494–520.

Rolt, L. T. C. 1959: *The London–Birmingham Motorway.* London: John Laing & Son Ltd.

Rolt, L. T. C. 1984 [1960]: *George and Robert Stephenson: The Railway Revolution*. Harmondsworth: Penguin.

Ross, K. 1995: *Fast Cars, Clean Bodies: Decolonization and the Reordering of French Culture*. London: MIT Press.

Sachs, W. 1992: *For Love of The Automobile: Looking Back into the History of our Desires*, Oxford: University of California Press.

Sampson, A. 1967: *Macmillan: A Study in Ambiguity*. London: Allen Lane/Penguin.

Shand, J. D. 1984: The Reichsautobahn: symbol for the Third Reich. *Journal of Contemporary History*, 19, 189–200.

Spence, B. 1959: Inaugural address of the president. *RIBA Journal*, 67(2), 36–8.

Spurrier, R. 1960: Road-style on the motorway. *The Architectural Review*, 128, 406–11.

Stamp, G. (ed.) 1986: *Sir Owen Williams 1890–1969*. London: The Architectural Association.

Thrift, N. 1996: *Spatial Formations*. London: Sage.

Thrift, N. 1999: Steps to an ecology of place. In D. Massey, J. Allen and P. Sarre (eds), *Human Geography Today*. Cambridge: Polity, 295–322.

The Times 1937: German trunk roads – British delegation invited – 1000 mile tour. 15 Sept., 14.

The Times 1959a: Trips to see motorway. 6 Nov., 6.

The Times 1959b: 5,000 cars an hour on motorway. 9 Nov., 10.

Urry, J. 2000: *Sociology Beyond Societies*. London: Routledge.

Yeomans, D. 2000: Collaborating with consulting engineers. In L. Campbell (ed.), *Twentieth-Century Architecture and its Histories*, London: Society of Architectural Historians of Great Britain, 125–52.

Chapter Seven

A New England: Landscape, Exhibition and Remaking Industrial Space in the 1930s

Denis Linehan

Dead King and the Underground

The designation of the term 'Special Areas' to depressed industrial districts in the north of England, South Wales and Scotland was the first attempt in Britain to devise a formal regional policy. For two weeks, between 30 January and 15 February 1936, these new spaces went on show in the London Underground.[1] Organized by the new Special Areas Commission, an exhibition in Charing Cross station, a short walk from the heart of government in Westminster, set out to 'tell the London public something of the story of the Special Areas, and perhaps indicate ways in which they can assist' (Special Areas Commission 1936a: 24). King George V, who had just died, was said to have given the exhibition his blessing, and the show was promoted as a typical example of the king's 'practical interest' in the welfare of the unemployed. According to *The Times*, the late king had 'wished to impress on Londoners that only a few hours train journey away were areas, formerly prosperous, where today there was widespread want of employment due to the decline in the principal industries'. It was seen as 'essential that the Nation at large should become conscious of the severe distress which existed within the boundaries of the Special Areas' (Anon. 1936a). The displays and the pamphlets given to the crowds from a kiosk at the exhibition's exit urged the citizens of London to go on their holidays to South Wales or to Cumberland and, when shopping, to 'take the trouble when making purchases to ask, where possible, for goods that were made in these areas' (Anon. 1936a). To ease the image of these regions as the home of heavy industry, care was taken to exhibit objects produced by light industry only. Presented beneath glass cases, these modern things included electric tools, health salts, baking powder, paint, wood preservatives and cigarettes.

Four large panels formed the background of the exhibition. The first, intended as an illustration of what was meant by Special Areas, was a composite photograph of derelict factories, slum cottages, smokeless chimneys and 'idle men'. The next panel depicted 'Recovery': new factories, ships loading cargo, lines of railway trucks converging on active collieries, and reclaimed industrial sites. The third panel depicted a panorama of activities: land settlement schemes, occupational clubs, keep fit centres for the wives of the unemployed and summer camps for schoolchildren. The last panel evoked wilderness, used pastoral images, and showed a number of beauty spots. To promote the event further, the first commissioner, Sir Malcolm Stewart, engaged the co-operation of the British Broadcasting Corporation, and on the opening day gave a short radio broadcast to the nation. Later, he estimated that approximately 80,000 people had visited the exhibition, and though it was meant to shut at 8 p.m. every day he was proud to report that 'on the closing day it was not possible to put up the barriers until about 9.45 p.m.' (Special Areas Commission 1936b: 25). For a time, it seemed, his success in the illuminated rooms beneath the surface of London cast a positive glow over the spaces of industrial dereliction that lay far beyond the furthest stop on the London Underground.

In Charing Cross station today Londoners rush for their evening trains, purchase their newspapers and furiously scan overhead timetables. Little of what might have been thought as people came to gaze at objects in an exhibition almost 70 years ago can be gleaned from this commotion. We can surmise, however, that to create a version of these regions and bring them to London, to be exhibited as if they were – despite the rhetoric of one nation – another country, was a move predicated on the assumption of a general unfamiliarity with and prejudice about these outer spaces recently renamed Special Areas. Stewart's reasoning was based on what he understood to be a wide distrust of South Wales and the industrial north, and the extensive and implied denigration of these areas in a range of official and popular surveys, press reports and hearsay. Anxious and sometimes apoplectic commentary on the state of the industrial regions, expressed through troubled assessments of community, landscape, body and soul, had in the preceding years formed into a coherent discourse about the depressed areas. If once the proud heartland of the Industrial Revolution, by the 1920s they had become increasingly marginal and contested places. Very quickly, as Rob Shields has argued about comparable peripheries, 'the image and stigma of their marginality became indistinguishable from any basic empirical identity they might once have had' (Shields 1991: 3). Thus this exhibition, with its rhetoric of progress, action and mobility, and its panoramas of happy children, suppressed a most modern concern, an anxiety about space. Curious commuters making their connections to the suburbs may well have been soothed by the joyful grins of land settlement workers, and impressed by statistics on economic growth and images of modern factories. As the work of the Commission progressed, emphasis was placed on the propaganda value of the state's interventions. The Commission paid for posters urging industrialists to locate in the depressed areas to be placed all over the London Underground and at international exhibitions in Brussels, Johannesburg and New York. Equally, photographs of new parks and playgrounds in Durham or South Wales, or of the unemployed engaged in vigorous exercise at physical culture exhibitions, would over the next few years become highly visible modes of intervention.

This chapter will consider the contexts of this exhibition, reflecting upon it not on the basis that it was a pivotal event, but in terms of how it engaged the geographies of modernity that characterize this period. The story at Charing Cross, projected enduring images of crisis and reconstruction that formed a grand narrative during this part of the twentieth century. The exhibition attempted to enrol the visitor into the politics of vision and order that informed the response of the state to the problem of reconstruction, and more particularly these depressed industrial regions, in the twentieth century. It follows that consideration of the dual processes of survey and exhibition will open up the way in which the modern and the region were intertwined. The chapter seeks to unfold elements of that process and unravel how the category of the modern and questions of geography were identified and mobilized. First, we will review some of the ways to survey and imagine the depressed industrial areas and new industrial districts around London. Secondly, we will examine the building of the 'new industrial city of tomorrow' in Gateshead in the late 1930s.

Envisioning a New England

Throughout the nineteenth and twentieth centuries, the representation of the industrial regions of Britain played a pivotal role in shaping national discourses about the consequences of modernity. From the late nineteenth century social investigators such as Booth and Rowntree had opened up the capacity of survey, but such forms of governmentality were ratcheted up in this period, as the institutions and capacity of social and political research increased. As economic crisis took hold and unemployment crept almost inexorably upwards, the regions became the focus of intensive investigation and speculation. A number of government surveys, notably those conducted by the Board of Trade during 1931 and the Ministry of Labour in 1934, engaged the problems of the depressed areas (Board of Trade 1932; Ministry of Labour 1934). These materials enrolled the region in a rapidly formed focus of national consideration, in which treatments of people and places wound into debates about the future of England. Just as, as Robert Gray has shown, debates about the factory in the nineteenth century served as a means of developing commentary about the condition of Britain, the industrial landscape became symbolic of national decline or progress. Hence, the condition of the industrial landscape and its inhabitants became a metaphor for expectations and anxieties about the pace and direction of modern transformation (Gray 1996). Eventually formalized into bounded regions which millions of British schoolchildren would learn about for their state examinations in geography, the region in question here is more dynamic. It was unquestionably a modern construct, the product of a continual process of myth-making, image-building, economic planning and political protest (Daniels 1994).

If the industrial surveys of the depressed areas represented a modernist narrative showing industry's fall from grace, the new industrial districts of the south and the Midlands were, by contrast, positively inscribed with modernist motifs. Enveloped in a discourse of modernization, the design, architecture and promotion of the new industrial districts that developed, around London especially, consciously sought to displace nineteenth-century codes of industry by seeking to envision progressive environments, distinct from the 'dereliction' of the contentious industrial landscapes of the north. The social construction of these industrial districts established them as new fields for Fordism. By the 1930s, sublime depictions of these industries emphasized their technological and progressive nature. While never officially surveyed, they were continually represented in the commercial and industrial press (Linehan 2003). Urban councils competed with one another to portray the most up-to-date and progressive industrial facilities. The new corporate landscapes envisioned in these advertisements were distinguished by typically modernist images of speed and lightness. The distinctive and self-conscious use of the language of modernity became both a signifier and a referent of the future (Vernon 1997).

Despite the reservations amongst dedicated modernists about the styling of this place, these new landscapes came to represent a modern environment for industrial development. This discursive process was embedded in the circulation of actual and symbolic capital. Investment was drawn into new industrial estates by their reduced cost base. At Slough the estate lay outside the London Wage Area, labour costs were lower, and the factories were rented rather than bought. The site was especially attractive to new industries, which could forgo large capital outlays. Soon these estates were believed to provide the recipe for success, in part by improving labour attachment and morale, and by claiming to facilitate uninterrupted, strike-free production flow in up-to-date factories. Cecil Day, industrial correspondent for the self-consciously progressive magazine *Business*, marvelled at the emergence of a new landscape of modernity from Ealing to

Perivale and on to Greenford, Uxbridge, Denham, Harefield and in a wide area stretching from the west London suburbs to Hertfordshire (Day 1936). At Welwyn Garden City, itself an example of progressive urban planning, the marketers draped a vision of scientific management over its industrial estate: 'The Factory Estate has been *scientifically planned* to provide all essential industrial services without obtruding upon the natural beauty of the surrounding country... Everything is modern light, bright, cheerful and efficient' (Welwyn Garden City 1936: 44). Another industrial estate at Elstree was praised for the fact that 'all the buildings are up to date and modern in every detail. All are brand new. They are designed to provide maximum facilities for economic operation' ('A Business Investigator' 1938: 32). In addition, the development of housing and amenities nearby was praised, as it would allow 'workers and executives to live within walking distance of the factories' ('A Business Investigator' 1938: 32). These kinds of facilities for industrial welfare were highly prized, as the managing director of Wolf & Co., manufacturing electrical tools at Ealing, explained:

> The workers can lead a much fuller life in the open surroundings of the district. A good example of what I mean is that they can walk to and from work. They can go home in the evening and develop a hobby such as gardening. The strain of the day's work can be thrown off quickly because directly they get out of the factory they can be in their homes which are themselves pleasantly situated. They don't have to make long journeys through the smog, roar, dust and fumes of the city to a place that is drab and uninteresting. (Day 1936: 21)

In the context of debates about Englishness, this is ambiguous territory (Matless 1998). The tone of George Orwell's essay 'England, your England', where he describes this landscape as part of emergent edge city of modern England, is dismissive and indefinite (Orwell 1941). When J. B. Priestley describes west London, he is reminded of California. Viscount Lymington was more forthright, suggesting the landscape had been destroyed, and that factories on the Great West Road would lead to a 'loss of nervous energy among those who have to travel to and fro from work' (Lymington 1938: 109). Most damning of all is John Betjeman's 1937 poem 'Slough'. Fully aware of the terror of Guernica, the poet conjured up bombers to demolish the town and 'blow it all to smithereens'. As John Carey has observed, such narratives reflected the negative tone of broader intellectual discourses on the effects of modernity (Carey 1992).

This intellectual constituency gives, however, only partial leverage into the condition of modern geographies. In his study of early twentieth-century American city planning, Fairfield has argued that Fordist ideas progressively pervaded the ideals of the city planning professions. He notes that 'comparing society to a complex factory and seeing themselves as social engineers, reform minded professionals adapted the principles of scientific management to the creation of an urban environment conducive to the accumulation of capital' (Fairfield 1994: 179). Such ideas are curiously neglected in British commentary. But idealized in west London is a place where the social and spatial regime is geared to the needs of the industrial capitalist, by providing housing for a labour reserve and attaching the worker to their employer through the provision of social and health services (Zukin 1991). The health centre at Slough was particularly singled out as a model for Britain's industrial future, and even had royal approval. When the king's mother, Queen Mary, went to visit she was so impressed she sent a book about doll's houses to be used in the estate's crèche. The modern it seemed, had become fixed in the spaces of the trading estate and Her Majesty sent 'every good wish for the prosperity and development of this splendid work'.[2]

Symbolized perhaps by its participation in the rhythmic routines of physical education organized by the Women's League of Health and Beauty, labour is also ordered in this landscape. The industrial bodies in the district were increasingly non-unionized, female and regarded as inherently passive (Scott 2000). This image of harmony and welfare contrasted with the northern industrial districts. Concerned about the effects of unionization in Lancashire on inward investment, the secretary to the Manchester Chamber of Commerce was particularly perturbed about this new 'virgin territory'. He argued that 'those who establish a new industry in the Thames Valley draw their labour from a rural district and their workpeople come fresh to industrial life without being permeated with all that is embraced in the term Trade Unionism' (Streat 1930). *The Times* also discussed the failure of trade union organizations here and suggested that the 'conflict of interest between employer and workpeople which is the *raison d'être* of the unions has scarcely arisen in industrial areas where employment is of a modern type, often in agreeable surroundings and has welfare amenities added'. It added: 'organizations which have espoused the theory of the class war and the irreconcilability of labour and capital do not come within the experience of the workers in these new areas' (Anon. 1935). The trade unionist Jack Gill warned the Labour Party of a similar problem: 'to develop a Trade Union outlook out of such mental material as this is a heart breaking task' (Gill 1936: 9).

The modernity of these new districts was accentuated through their products: disposable razor blades, telephones, radios, washing machines and vacuum cleaners, produced by consumer-driven and increasingly Americanized companies such as Schweppes, Johnson & Johnson, Mars, and Black & Decker. It became commonplace to expose the modernity of the production process to public scrutiny, representing it as a marvellous event, as a mark of human ingenuity and invention. Within a broad range of literatures in economics and engineering, questions of efficiency and equilibrium linked the practice of industry and the spaces in which it could take place. While economic historians caution about how much of this was achieved, the ideal was a powerful discourse. Reading British commercial newspapers and engineering journals, we can see bodies, machines, factory spaces, and locations encased in a scalable geography which had to be orderly and efficient (figure 7.1). Sublime scripting of factory work, invoking motifs of wonder, magic and awe, often characterized these representations. Factory visits, a kind of industrial tourism, a holiday with modernity, were commonplace. When the Carreras cigarette factory opened on Hampstead Road in north London press reports emphasized the modernity of the building, where, in addition to an exotic exterior, 'no expense was spared on using every scientific means to improve and maintain the production of the perfect cigarette'. Even the 'weather' inside was controlled automatically:

> Whether doors or windows are open or shut; whether London is in the grip of a severe frost; or sweltering under a summer sun; or enveloped in the mugginess of a November fog – by means of a scientific apparatus, the incoming air is washed with water and its humidity and temperature adjusted so that everyday is a perfect day in Carreras's new Factory. (Anon. 1928)

> Pride in the process of production is married with a sense of its newness and departure from tradition. Everywhere, nature was re-ordered and regulated, such as at the Ford plant at Dagenham where 1000 tons of London refuse was used daily to fire the furnaces and the temperature of the plant could be controlled by a machine that automatically opened and closed 'six miles' of windows. (Anon. 1932)

These representations of the new industrial districts were in stark and dramatic contrast to the ways in which the depressed areas were portrayed; what the Communist MP Arthur Greenwood referred to grimly as 'a vast literature of social discouragement that brought in its train its own Nemesis' (Greenwood 1935). Set beside the narratives of the distressed areas, these stories of success seemed to exacerbate the deterioration of the older industrial districts. Government surveys were characterized by themes of degeneration. Indeed, the landscapes rendered in these reports by the Board of Trade and the Ministry of Labour provide a textbook example of what Marshall Berman noted as Marx's fascination with the melting metaphor in his descriptions of the chaos caused by rapacious capitalism (Berman 1988). The reports are inflected with a sense of collapse, disintegration and social, economic and moral breakdown (Linehan 2000). State surveys of the depressed areas encouraged the superimposition of physical and psychological space, a factor that was implicated in meanings associated

Planning the lay-out of your factory

Manufacturing processes to-day demand complete freedom in plant lay-out. To achieve this, first plan your lay-out in its logical sequence. Then design your factory around the plant, so that it functions efficiently and without restriction. Remember that a well-designed factory is very like a well-tailored suit. It exactly fits its purpose, yet permits unrestricted movement.

COMMERCIAL STRUCTURES, builders of CS planned factories, provide a complete service that takes care of all lay-out problems. CS Engineers, working in close co-operation with the manufacturer, evolve the plant lay-out that will give the greatest possible efficiency. Qualified architects prepare plans for a building designed exactly for its purpose. All building work—steel, concrete, heating, etc.—is carried out by Commercial Structures, making possible a big saving in cost and time. If you are considering moving to a new factory, or extending your existing premises, you will find CS ideas helpful. Building suggestions and estimates will gladly be submitted without obligation. Write or 'phone . . .

Commercial Structures Ltd

Figure 7.1. Commercial Geographies
Source: Business (Mar. 1939).

with economic 'depression'. In his survey on the north-east, the investigator Euan Wallace noted that 'the psychological effects of depression are not confined to particular spots but permeate the whole region . . . impossible to avoid the strong general impression that the area as a whole is losing hope' (Wallace 1934: 84). Initially, Stewart himself sustained such imagery, arguing that it 'is beyond the power of man to turn the tide of trade whence it is ebbed; derelict areas are the wastes and scars left behind' (Stewart 1935: 286) The condition of the industrial landscapes became part of the wider dimension of post-war crisis and were permeated with 'a distinctly modern sense of a whole world giving away, a permanent impermanence' (Gregory 1994: 215). Often portrayed as dark, enclosed or isolated, the landscape could picture the condition of the environments of the depressed areas as the very antithesis of a progressive and healthy nation. Appalled by their condition, *The Times* protested that 'thousands and thousands of men and boys are sinking through idleness into a miserable inertia, and families and communities are involved in a steadily accentuated hope-lessness. The sound mind, the cheerful spirit, are well-nigh impossible in the environment of the gloomier part of the Special Areas' (Anon. 1936c).

The political commentary on the 'distressed areas' never escaped the modernist ghost of planned and regular order. Paradoxically, the solution for this modern disaster was modernism. Order needed to be put back in. The debate about how that was to be achieved was highly contested in this period (Ritschel 1997). The maverick Labour MP Andrew Maclaren criticized 'the hollow farce of further investigations' and lampooned the government's plans for regeneration:

> Even if we did tidy up these areas, the prospective innovators of new indus-tries would still be apprehensive, because of the Socialist and Labour views of the working man in that particular area. So much for the prospect of a 'tidied area' with Socialists still living in it – apparently it would be necessary to remove agitators, Socialists and trade-unions as well as slag-heaps. (Maclaren 1935: 4)

As a key figure in this policy environment the first commissioner, Malcolm Stewart, was key in mobilizing a progressive approach. Modern-izing influences are reflected throughout his background and outlook. As director of the London Brick Company, and Portland Cement, he was a consistent modernizer, in the vanguard of industrialists in the London area to introduce welfare, pensions and holiday schemes. His company's stand at the Building Exhibition at Olympia in 1936 was designed by the architect Julian Leathart, who by then had completed his stylish and quintessential modernist cinema in Dreamland at the seaside resort of Margate. The exhibition Stewart organized at Charing Cross had the mark of modernity all over it. Stewart saw the event as a 'valuable example of the application of Modern publicity methods to a National Problem' (Special Areas Commission 1936a: 25). Charing Cross itself was one of the most chic exhibition spaces in London, and the company he commissioned to organize it, Pritchard & Wood, was more fashionable still, with connections to the Isobar and other centres of metropolitan modernism in the city. Stewart was also the joint employee chairman of the National Industrial Alliance, an organization dedicated to industrial planning and rational labour relations. Every month the association's journal published a section on the 'New Industrial England' outlining successes in industrial production and welfare. This progressive outlook Stewart made material in a model factory village in Bedfordshire, named after his family. The plant's nine tall chimneys, spelling S-T-E-W-A-R-T-B-Y, can be still be seen for miles by travellers on the rail line from London St Pancras to the east Midlands. As he was chauffeured back and forth from the Ministry of Labour to his home, he saw through the windows of his car the possibilities of modern industrialization laid out before him.

The Industrial City of Tomorrow

Industrial estates were planned for all the Special Areas, but the trading estate at Team Valley was singled out as the jewel in the Commission's crown. The site was in north-east England, close to Gateshead and Newcastle. In June 1937 the new king, George VI, and other visitors to the British Industrial Fair at Earl's Court in London, viewed a 50 ft^2 model of the proposed development. While, a year earlier, Londoners could gaze at modern products in glass cases in Charing Cross, now they could survey an industrial future made up 'miniature factories, roads, railways and gardens' (Anon. 1937b). Billed as the 'finest trading estate in Europe', the model toured the country that summer, travelling to Bristol, Birmingham and Nottingham. The site was regarded as suitable for light industry, and in his description Stewart tripped off nearly all the codes associated with modern industrial planning. Connectivity and speed were at a premium. The estate would be near sources of labour, rail and road links and was regarded as 'the natural centre from which industry can supply both the local and national markets' (Special Areas Commission 1937a: 69). In common with locations in the south, the welfare of the workers would be assured by 'open spaces, playing fields, canteens and other amenities in order to create pleasant and healthy conditions for the work-people employed thereon' (Special Areas Commission 1937a: 69). The location was applauded as it was away from the heavily industrialized landscape in the region, and plans were made to preserve the 'especially attractive' rural surroundings through the purchase of surrounding lands (Special Areas Commission 1937a: 69). The whole estate, he wrote, would be based on 'a model plan based on modern and attractive lines' and the buildings would be 'throughout homogeneous in design and up to the highest standards of modern practice' (Special Areas Commission 1937a: 69).

These plans were involved in a process of spatial inflection, whereby models of industrial space, organization and architecture developed in Trafford Park in Manchester, and especially Slough and other industrial estates around London, were replicated.[3] In arriving at this formulation Stewart was informed by the consulting engineering firm Alex Gibb & Partners, and by K. C. Appleyard, whom he had chosen to be chair of North-Eastern Trading Estates Ltd. In Appleyard he found a visionary entrepreneur, remarkably imbued with a sense of his modernist mission:

> Having got the Estate, my Colleagues and I are then of the opinion that we have the opportunity of showing what can be done in the way of Industrial Planning on a small scale in the most modern fashion, and it is therefore our intention to lay out the Estate in such a way that it will not only be an efficient Industrial Unit but will also be as beautiful as it can be . . . Moreover . . . it is necessary to create our Estate in pleasant surroundings which will give people the feeling they are going into a Rural atmosphere rather than into the depressing atmosphere always associated with the Iron and Steel and Heavy Engineering industries, and perhaps rather particularly justified by the dere-lict appearance of the River Tyne generally.[4]

The anxieties related to the reconditioning of the environment went hand in hand with concerns for the reconditioning of labour. Appleyard acknow-ledged that 'there is a very much greater tendency at the present time to consider the conditions under which workers live than has been the case in the past. Their moral and physical well-being is now receiving very much more attention, and by many firms this is considered an important aspect of their business.'[5] Stewart had hoped that the exhibition at Charing Cross 'would catch the eye of many industrialists who were thinking of starting new industries', and argued that the position of the estate from an adver-tising standpoint could not be ignored. He thought it 'desirable that the situation was such that it could be readily seen from both the main north and south railway lines and from a main arterial road' (Anon. 1936a). These considerations underline the significance of Team Valley as an exhibition space. The industrial estate emerged as a space that was an intervention in unstable economies, and was an attempt to promote confidence in the region. It supported the Commission's approach to landscaping and improvement, and was underwritten by a politicized ra-tionale which involved the renegotiation of existing representations of the areas. Much as its policies were attempts to rework the landscape and ameliorate social conditions, they were also attempts to write off signs of the past and signal a different vision of the future.

These concerns were certainly in the mind of the architect W. G. Holford when he won the contract to design the estate in 1936 (Cherry and Leith 1993; Holford 1939). Holford laid out the estate with a rectangular grid, forming factory blocks ranging from 19 to 25 acres with factories facing the roads, the larger ones on the main roads, and the smaller ones on the sidestreets. The centre of each quadrangle was planted and laid out with paths and benches, and many of the factory groups had their own canteen, cycle sheds, telephone booths, and bus stops. The administrative block was in the centre of the estate, and was originally intended to include the estate offices, a restaurant, bank, and post office, and police, first-aid and fire stations. Not long after the foundations for Team Valley were built, Stewart resigned as commissioner, apparently for personal reasons, but his frustrations with protracted and often obstructive government ministries had taken their toll. In two years, he had become a convert to planning and hence a more persistent thorn in their side. His final report presented a radical critique of the development of London, and stimulated the establishment of the Royal Commission into the Geographical Distribution of the Industrial Population.

At this point, however, the work and the myth-making were well under way. The next commissioner, George Gillett, continued to represent Team Valley in progressive language. In the *Architectural Journal*, he was eager to illustrate its modernity, and argued that 'a trading estate can be planned from the outset in such a way to produce harmony both between its separate units and between the whole estate and its surroundings' (Gillett 1937: 16). By September 1937, he was enthusing further about his new 'industrial city'. Displaying a commitment to a rational, planned future, he anticipated that the estate would be a model for both industrial development and environmental reconstruction, 'enhancing the value and appearance of the whole district'. Paying attention to 'aesthetic considerations', he gave an assurance that the 'consulting architect has given great care and attention to the lay-out of factory blocks, and the design has met with the hearty approval of incoming tenants and all visitors to the Estate'. Gillett told the public that 'the factories themselves blend pleasantly into the landscape' and marvelled at 'what can now be seen as a result of proper industrial planning on model lines based on a bold conception' (Special Areas Commission 1937b: 73–4).

The estate company itself continually reproduced it own myth as modern. Advertisements for the estate reinforced notions of connectivity and speed and drew attention to the proximity of the site to the Great North Road and the Great Northern railway line (see figure 7.2) When K. C. Appleyard came to write the story of the estate for *The Times*, he framed it in terms of a modern miracle: 'where two years ago there was an area largely consisting of marsh, rubbish dumps, and smallholdings, today exists a magnificently planned lay-out for industrial enterprise, where light industries, thriving and successful, have sprung up literally by the thousand' (Appleyard 1939a). While this material is recognizably promotional, the allusions to the modern remain central to the rational and lived interpretation of the estate. Modernity is presented as more than simply being 'up to date'; rather, as Griselda Pollock argues, it is 'a matter of representation and major myths' (Pollock 2000: 51). The key markers of this mythic territory are order, connectivity, light industry, modern labour and success. In an encounter with the estate's geography and his own modernity, Appleyard had a vision:

> I can see forty thousand men, women and juveniles coming to work from around the area in motor coaches and railway carriages on the network which the transport authorities have built up with the Estate at its centre. They will enjoy work because they will be employed in airy well lighted factories, and they will remain fit because on our playing fields, in our recreation centre, and amidst our gardens and trees, they will find health and contentment.

The landscape was exploited to represent ideas of social order. Sounding like a supercharged Soviet commissar, Appleyard marvelled how, during 'all the time the work of selling Team Valley to the world has been going on, there has been a small army of men, civil engineers, architects, contractors, builders and electricians, untiring and enthusiastic, creating an ideal city within 700 acres of pasture land'. He was pleased to report that 'the once winding river Team is straightened and controlled' and that 'nine pit heaps, those ugly landmarks of our district have been demolished' (Appleyard 1939b). As a consequence, he had told the Newcastle branch of chartered surveyors that 'the main avenue would make the Great West Road a mere pup by comparison' (Anon. 1937c). This hyperbole was sustained by another of the estate directors, H. A. Sisson, in an address to the Wallsend Rotary Club, when he declared that 'not only has the North East been put on the map, but we have made a new map to put it on!' (Anon. 1937a). The discourse of modernization which pervaded the planning and organization of the estate would also be reflected in the corporate culture promoted within it. Sigmund Pumps BG Ltd. fully embraced the culture of Team Valley, linking its own future and its labour force with the modernist architecture of the estate (figure 7.3). The company instituted a range of welfare and training programmes, and through its apprentice scheme sought to offer progressive training in engineering, take care of the boys' health and fitness, and vowed to develop the 'character and citizenship during the boys' most receptive years' (Sigmund Pumps Ltd., 1942). Press coverage of the estate played on such rhetoric and imagery and the spectacle of the 'ultra-modern layout' of the estate was continuously marvelled at and imagined (Anon. 1936d). Sublime photographic essays on the

COMPLETE COMMUNICATIONS

Eight miles of Estate railway bring the L.N.E.R. main line to the very factory door. Sixteen miles of wide Estate roads join each factory with the Great North Road and other arteries. London can be reached overnight by rail or road. The wharves of Tyneport are a mile from the Estate.

IMPORTANT SERVICES

13 well-qualified business men of 13 different professions are at the disposal of tenants. 21 experts on Architecture, Engineering, Lighting, Heating, Water and Effluent make no charge for their advice. The Estate is equipped with canteens and playing fields for workers, restaurants. club. committee rooms, etc., for executives.

TEAM VALLEY
TRADING ESTATE

To-day's Industrial City of To-morrow

NORTH EASTERN TRADING ESTATES, LTD. TEAM VALLEY. GATESHEAD-ON-TYNE
Tel: Low-Fell 76071

Figure 7.2. Exhibiting Team Valley
Source: The Times Trade and Engineering (Apr. 1938) (detail). Reproduced courtesy of the Urban Regeneration Agency (known as English Partnerships)

Figure 7.3. Modern Boys
Source: Sigmund Pumps Ltd. 1943, *Looking Forward*, Gateshead.

floodlit factories were commonplace and the *Newcastle Journal* even depicted the site with an infra-red photograph.

The estate was grasped by progressive groups such as Political and Economic Planning and the Labour Party as an example of regional and economic planning, a measure the national government was hesitant to legitimate. While it was acceptable that Team Valley could be publicized as a government initiative to deal with unemployment, like other initiatives undertaken by the Special Areas Commission, the policy rationale was officially represented as experimental. It was to lead by example to private enterprise, not to further public intervention. However, by the late 1930s the tide of modern planning could not be turned. When J. C. Alfield, the President of the Town Planning Institute, visited Team Valley in 1938 he announced he that 'found it infinitely more pleasant, more spacious, clean and refreshing that the speculatively created residential area through which I had passed in order to reach it' (Alfield 1938: 12). In a way that would mirror the Labour Party's appropriation of the symbolism of this industrial space in the post-war period, the trade union leader and MP J. R. Clynes spoke passionately about the 'astonishing' potential of industrial planning. He told the *North Mail and Newcastle Daily Chronicle* that 'this is very much like turning a rubbish heap into an industrial garden'. Team Valley would be an example not just to local industrialists, but 'to the whole country, and if other national projects could be planned in this way, the world would be a better place in which to live' (Anon. 1938).

There is a range of issues wrapped up in the discourse of this place and its landscape. The modernity of the estate allowed workers to be rescripted. Depictions of passive bodies travelling to Team Valley in modern networks, and engaging in rational recreation, sought to represent labour as adaptable and apolitical. Team Valley demonstrated unequivocally that an industrial landscape is also a moral order (Zukin 1991). Ensuring in addition that local politicians, the majority members of the Labour Party, were excluded from the management of the site was also part of the strategy to depoliticize the space. These were reconditioned people in reconstructed landscapes. While Maclaren had lampooned the state's plans for reconstruction, it did, initially at least, become possible to remove agitators, socialists and trade unions, as well as slag-heaps. It was noted by many that when the wire mesh that covered the River Team eventually bloomed with roses, none of them was red. Hence, the injection of capital into the region was a way of reordering both landscape and society, and an attempt to erase previous histories. The estate thus not only operated as a location for factories, but was embedded in a mental landscape and an ideological map.

After

Following the Second World War, the modernity of this site and the political significance of this landscape did not disappear. The 'new map', the 'industrial city', and the skating rinks and tennis courts promised by the visionaries of the 1930s did not exactly materialize. While there was a lot of open parkland and a few playing fields, the dream of rational recreation remained a dream. Despite these limitations, the mobilization of this landscape by the post-war Labour government continued to enrol it in a modernization project. Team Valley and the other sites were used to bolster the modernity of the Labour government. The exhibition quality of the 'Development Areas' – the new term adopted for these industrial regions in 1945 – was not lost. The difference in some ways is measured in terms of scale. By 1947, the Labour Party would claim that it was responsible for providing employment for 220,000 people in 925 new factories, and that these interventions were 'rebalancing' the regional economies (Labour Party 1947). For the Chancellor of the Exchequer, Hugh Dalton, the policy was 'a striking example of what intelligent Socialist Planning can do to repair the blunders and the social devastations of an uncontrolled capitalist economy' (Labour Party 1949). As Jeremy Seabrook has noted, the 'emancipatory power of landscape has always been a powerful, somewhat mystical impulse in socialist thought' (Seabrook 1985: 41). This chapter has queried the manner in which questions of the modern landscape and questions of geography were enrolled in the reimagining and reconstruction of Britain. Today, Antony Gormley's monumental sculpture *Angel of the North* guides travellers towards Team Valley and Gateshead. The sublime quality of the giant figure, whose wing span is, we are told, almost the same as that of a jumbo jet, revisits the fact that this apparently peripheral site has been a significant place in the history of British modernity. The shift from the 'industrial city of tomorrow' to an angel in the landscape may not seem progressive to some, but it illustrates the enduring power of narratives of the modern and of reconstruction in the making of a New England.

NOTES

1   The areas designated under this scheme were the greater part of Durham, the Tyneside and Haltwhistle areas in Northumberland, West Cumberland, the counties of Glamorgan and Monmouth, a part of Brecon and the borough of Pembroke.
2   Letter to Lady [Nancy] Astor from Cynthia Colville, 30 Apr. 1937. Public Record Office (PRO) ED 113/77.
3   Report on sites for trading estates prepared by Alex Gibb & Partners, PRO BT 104/31.
4   Letter to the Commissioner from K. C. Appleyard, 18 Aug. 1936 PRO BT 104/23.
5   Ibid.

REFERENCES

'A Business Investigator', 1938: Would a factory at this 'strategic point' estate smooth out your management problems? *Business*, May, 31–6.

Alfield, J. C. 1938: Presidential address. *Journal of the Town Planning Institute*, 25, 4–12.

Anon. 1928: An announcement by Mr Bernhard Baron on the opening of Carreras New Wonder Factory. Advertisement in *The Times*, 6 Nov.

Anon. 1932: New Ford works at Dagenham: an achievement in mass production. *The Times*, 25 June.

Anon. 1935: Editorial. *The Times*, 21 Aug.

Anon. 1936a: The Special Areas: an exhibition for Londoners. *The Times*, 31 Jan.

Anon. 1936b: Helping the Special Areas progress illustrated at exhibition. *The Times*, 30 Jan.

Anon. 1936c: The distressed areas. *The Times*, 14 Feb.

Anon. 1936d: Gateshead Trading Estate. *The Newcastle Journal*, 25 Nov.

Anon. 1937a: Trading estate. *North Mail and Newcastle Daily Chronicle*, 7 Jan.

Anon. 1937b: Untitled article. *The National Builder*, Feb., 255.

Anon. 1937c: Tyne Trading Estate. *North Mail and Newcastle Daily Chronicle*, 14 Oct.

Anon. 1938: Vast play centre on trade estate. *North Mail and Newcastle Daily Chronicle*, 2 Mar.

Appleyard, K. C. 1939a: Team Valley Estate. *The Times Trade and Industrial*, 1 June, Special Areas number, 8–9.

Appleyard, K. C. 1939b: The Team Valley Estate and its success. Public Record Office BT 104/28.

Berman, M. 1988: *All That Is Solid Melts Into Air*. Harmondsworth: Penguin.

Board of Trade 1932: *The Lancashire Areas* (51.196), *Merseyside* (51.193), *The North East Coast Area* (51.194) *Scotland* (51.191), *South Wales* (51.192). London: HMSO.

Carey, J. 1992: *The Intellectuals and the Masses*. London: Faber & Faber.

Cherry, G. and P. Leith 1993: *Holford: A Study in Architecture, Planning and Civic Design*. London: Mansell.

Daniels, S. 1994: Inventing the East Midlands. *East Midland Geographer*, 17, 3–11.

Day, C. E. 1936: Western London . . . is industry's new 'clean air' area. Factory Site Survey 10, *Business*, June, 21–2.

Fairfield, J. 1994: The scientific management of urban space: professional city planning and the legacy of progressive reform. *Journal of Urban History*, 20, 179–204.

Gill, J. 1936: The new industrial Britain. *Labour*, Sept., 9

Gillett, G. 1937: Team Valley Trading Estate. *Architect's Journal*, May, 20.

Gray, R. 1996: *The Factory Question and Industrial England 1830–1860*. Cambridge: Cambridge University Press.

Greenwood, A. 1935: *Hansard*, 23 July, col. 1686.

Gregory, D. 1994: *Geographical Imaginations*. Oxford: Blackwell.

Holford, G. 1939: Location and design of trading estates. *Journal of the Town Planning Institute*, 25, 151–63.

Labour Party 1947: *Bringing Work to the Workers*. Labour Press Service, 12 Oct.

Labour Party 1949: *Re-equipping Britain: Bringing New Life and Hope to the Old Distressed Areas*. London: Labour Party.

Linehan, D. 2000: An archaeology of dereliction. *Journal of Historical Geography*, 26, 99–111.

Linehan, D. 2003: Regional survey and the economic geographies of Britain, 1930–1939. *Transactions Institute of British Geographers*, 28, 96–122.

Lymington, V. 1938: *Famine in England*. London: The Right Book Club.

Maclaren, A. 1935: *The Truth about the Distressed Areas*. London: The English League for the Taxation of Land Values.

Matless, D. 1998: *Landscape and Englishness*. London: Reaktion Books.

Ministry of Labour 1934: *Reports of the Investigation into the Industrial Conditions in Certain Depressed Areas*. Cmd 4728. London: HMSO.

Orwell, G. 1941: England your England. In *Inside the Whale and Other Essays*. Harmondsworth: Penguin, 63–90.

Pollock, G. 2000: Modernity and the spaces of femininity. In *Vision and Difference: Femininity, Feminism and the Histories of Art*. London: Routledge, 50–90.

Ritschel, D. 1997: *The Politics of Planning*. Oxford: Oxford University Press.

Scott, P. 2000: Women, other 'fresh' workers and the new manufacturing workforce of inter-war Britain. *International Review of Social History*, 45, 449–74.

Seabrook, J. 1985: *Landscapes of Poverty*. Oxford: Basil Blackwell.

Shields, R. 1991: *Places on the Margin*. London: Routledge.

Sigmund Pumps Ltd. 1942: *Looking Forward*. Gateshead: Sigmund Pumps Ltd.

Special Areas Commission 1936a: *Helping the Special Areas*. London: Special Areas Commission.

Special Areas Commission 1936b: *Second Report of the Commissioner for the Special Areas (England and Wales)*. Cmd 5090. London: HMSO.

Special Areas Commission 1937a: *Third Report of the Commissioner for the Special Areas (England and Wales)*. Cmd 5303. London: HMSO.

Special Areas Commission 1937b: *Report of the Commissioner for the Special Areas in England and Wales for the Year Ended September*. Cmd 5595. London: HMSO.

Stewart, M. 1935: The Special Areas. *The Ministry of Labour Gazette*, Aug., 286.

Streat, R. 1930: Memorandum on the encouragement of new companies in Lancashire. Minutes of the Lancashire Industrial, 29 Nov., Manchester Central Library, M199.

Vernon, J. 1997: The mirage of modernity. *Social History*, 22, 208–15.

Wallace, D. 1934: Durham and Tyneside. In *Report of the Investigations into the Industrial Conditions of the Depressed Areas*. Cmd. 4728. London: HMSO.

Welwyn Garden City 1936: A factory estate planned for Industry. Advertisement in *Business*, Sept., 44.

Zukin, S. 1991: *Landscapes of Power: From Detroit to Disney World*. Los Angeles: University of California Press.

Chapter Eight

A Man's World? Masculinity and Metropolitan Modernity at Simpson Piccadilly

Bronwen Edwards

On 29 April 1936 a crowd gathered in London's Piccadilly in front of a striking new shop. Somewhat inauspiciously for a future modernist icon, it was sandwiched between a Lyons tea shop and the National Provincial Bank, on the site of the old Royal Geological Museum. Many of the crowd were there to catch a glimpse of Sir Malcolm Campbell, holder of the world land speed record, who was to open the Simpson Piccadilly building. They were also curious to see a new retail experiment, claiming to be the first men's department store, hailed in the *Evening Gazette* as 'a new bright spot in the lives of men' (29 April 1936).[1] This was an important moment in the history of the West End, when a new model of fashionable masculinity collided with cutting-edge architectural design at 202 Piccadilly, disrupting and reconfiguring established shopping routes and practices. For this 'bright spot' lay at in the heart of London's West End shopping district, and created a new intersection between two very different retail routes running through it.

The following study of Simpson Piccadilly (more usually known as just 'Simpsons') locates fashion in its metropolitan setting, and contributes to a developing body of work in urban history concerned with the significance of gendered consumption practices (Breward 1999; Domosh 1996; Nead 2000; Rappaport 2000). The creation of the 'Simpsons man', defined by the relationship he had with fashion, consumption, leisure and the city, helped to legitimize a new kind of English masculinity. The use of modernist architecture to house this type provides insights into a particular version of modernity prevalent in 1930s London. Leading artists and designers were drawn together to work on a project which was unconnected with the high social ideals often associated with international modernism, but was instead in the business of selling fashionable clothes. This brand of modernity was built into the consumption practices and the very fabric of the building, making a permanent impact on the West End.

Recent historical work on consumption practices in the West End has largely focused on the Victorian and Edwardian periods. One result of this particular approach to the relationship between gender identity and the spaces of the West End has been the production of a composite map with quite distinct shopping areas and shopping routes for men and women. Erica Rappaport (2000) tells the story of the establishment of a feminine stronghold in the West End. She argues that, by the outbreak of the First World War, a new kind of feminine consumption was flourishing, centred on the major department stores along Regent Street and Oxford Street. Christopher Breward (1999) has identified a coexistent group of fashionable male consumers, hidden from existing histories, but occupying an older geography of masculine consumption threading through the back streets of the West End. By 1936 London's male consumers were becoming less 'hidden', and were to be found grabbing the headlines and disrupting the established gendered geographical order. Simpsons' precise location on this shopping map was crucial for the store's identity. Simpsons also had a transformative effect on the West End as a whole, as a new kind of intersection between the routes of masculine and feminine consumption and as a pioneer of new forms of masculine consumption, combining modern retail methods and modernist style. The building itself was the result of a unique collaboration between Alexander Simpson, rising star of menswear manufacturing already famous for his 'Daks' trousers, and Joseph Emberton, modernist architect with special expertise in 'leisure architecture': shops, exhibitions and funfairs. Together they created a building tailored to fit a new kind of Englishman: the Simpsons man.

Simpsons: A West End Store

The 1930s can be seen as a mature period in the history of London department stores, when anything could be bought conveniently in one place if the customer so desired. Simpsons was a department store rather than simply a large men's outfitter, its aim 'to bring together, into a beautiful and convenient setting, all the best things which are made for men' (*Evening Gazette*, 30 April 1936). Simpsons' extensive clothing stock ranged from full dress suits to sportswear, but the store went much further, providing everything a man might wish for. The lower ground floor alone was designed to house a barber's shop, soda fountain, gun shop, shoe shop, chemist, florist, fishing shop, wine and spirit shop, luggage shop, snack bar, dog shop, sports shop, tobacconists, gift shop, saddlery shop, and theatre and travel agents (Simpson–Emberton correspondence 2 September 1935). During the opening months there was even an aviation department exhibiting full-sized aeroplanes.

Such a shop might be expected to operate independently of its environment, but contemporary street photographs, trade directories, tourist guides and descriptions in literature and magazines all suggest that the shopping trip was not easily contained within a single department store. West End shopping worked within the wider street network, and was characterized by window shopping, browsing and comparing the goods of different stores. As Rappaport has shown (2000), this milieu was also coloured by the tea-rooms, restaurants, theatres and clubs producing a composite experience of being 'in town'. The major stores drew heavily on their position in the West End in the operation of their businesses. Simpsons was keen to establish links with the city beyond its windows. It housed a successful theatre booking service, and even exploring the possibility of a tunnel linking it with the neighbouring Lyons tea house. The very name 'Simpson Piccadilly' established an immediate and permanent relationship with the street, which was emphasized in every advertisement, every newspaper article, even every time someone spoke of the store. This focus did more than simply help customers to find the store; it placed the location at the centre of the store's character.

The precise nature of the West End requires unravelling. It was a fluid place, whose character changed during the course of the day. It was for many the destination of a journey, the site for a 'day in town' for the suburban and provincial populations, as well as for tourists. It was also a network of routes, characterized by an ever-increasing visual and aural cacophony created by the constant movement of vehicular and pedestrian traffic through its streets. The development of the transport system of tube, buses, trains and taxis was key to the functioning of the West End. The directors of Simpsons were keen to secure a bus stop outside the Piccadilly entrance, and promoted the store as somewhere that could be visited by businessmen on their way home from work. The West End needs to be read as a web of streets, and as multiple and fragmented in character. Even the tight cluster of streets in the centre of the West End, around Oxford Street, Regent Street and Piccadilly, actually comprised a series of overlapping shopping routes, subtly differentiated from each other in terms of the gender and status of the intended customer, meeting each other at particular points. Simpsons' positioning within this web was crucial for its identity, and its opening was to disrupt and reconfigure the shopping routes and practices of the West End. Simpsons had two entrances: one on Piccadilly, close to Piccadilly Circus, and the other on Jermyn Street. The shop itself was an intersection between two very different retail routes, mediating tensions between old and new approaches to men's retailing, and indeed to ideas of modern masculinity itself (figure 8.1).

Piccadilly Circus was an important nexus in the West End, at the junction of several major thoroughfares: Shaftesbury Avenue, Haymarket, Regent Street and Piccadilly. Photographs and postcards from the 1930s show it as a busy, even chaotic place. A popular saying from the period, reproduced on postcards, suggested that you would see the whole world pass before you if you stood there long enough. Contemporary social commentators certainly voiced anxieties about the broad social mix. Simpsons was in danger of an influx of customers lacking the desired class profile, and the store's directors were certainly preoccupied with preserving its exclusive image. But the proximity to Piccadilly Circus also helped to attract a sufficient volume of customers to fill a store of Simpsons' proportions, vital for the success of the venture. The store was therefore eager to associate itself with this important local landmark, using the statue

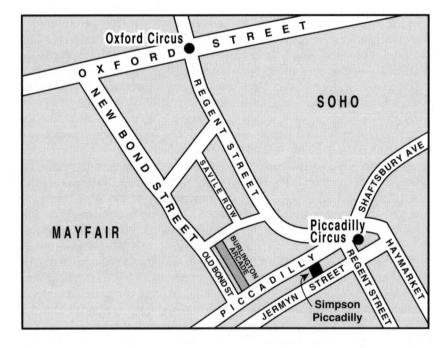

Figure 8.1. Simpson Piccadilly's position in the West End placed it at an intersection between the traditional masculine shopping route running through Savile Row and Burlington Arcade to Jermyn Street, and the more feminized West End of Oxford Street and Regent Street

of Eros in Piccadilly Circus in sales advertisements and store guides for many years, claiming it as a signpost to the store.

Piccadilly Circus was also the start of the West End's main shopping artery: a plush trail of famously showy department stores running from Piccadilly Circus, up Regent Street and along Oxford Street. This route was well established as the site of feminine consumption and leisure by the 1930s. These stores were characterized by dramatic architectural statement, and by the opulence and novelty of display methods. By placing Simpsons' eye-catching main entrance on Piccadilly the business addressed this route. In contrast, Simpsons' more discreet back entrance connected with the well-established tailors, shirtmakers, bootmakers, tobacconists and clubs on Jermyn Street, part of the West End route of traditional masculine consumption that included Savile Row and the Burlington Arcade. These streets were the heartlands of traditional masculine consumption, and were lauded by contemporary accounts in guidebooks and newspapers as one of the few remaining spaces in the capital where time stood still. Modernity was spurned by the continued use of old-fashioned shop fronts, window displays and retail methods. The streets were significantly slightly off the beaten track, and a certain amount of privileged urban knowledge was required to find and use them. By straddling the two routes, Simpsons was defining itself as part of a new breed of 'man's shop', which included the innovative chain of Austin Reed, with which important parallels can be drawn. These shops were increasingly staking a claim for themselves in the traditionally feminine shopping route.

Although Simpsons occupied exactly the same long, narrow footprint as the Royal Geological Museum, which previously stood on the site, the new building was reoriented. This meant that 202 Piccadilly now functioned in a completely different way in the West End street network. The Piccadilly palazzo stone frontage of the old museum, although its grander façade, had had no street access; the building's entrance was via the altogether quieter brick and stone frontage in Jermyn Street. While the museum had turned its back on Piccadilly, communing rather with the traditional male preserve of Jermyn Street, Simpsons embraced what it found on the grander, busier street, without wanting to sever its links with the quieter one. The creation of a second access point to the building was also crucial, for it enabled the store's own internal circulation routes to tap directly into the street network. Simpsons' two entrances meant that it provided a direct cut-through from Piccadilly to Jermyn Street, functioning like an arcade. But this was not without its risks: male consumers, popularly regarded as rather conservative in their habits, had to be attracted into this new shopping environment.

'An expression in every way of the modern spirit': Modernism and Masculinity at 202 Piccadilly

At Simpson Piccadilly modern design was used to indicate a modern clientele, and to legitimize a new way of selling menswear. The *Scotsman* (4 May 1936) made this connection in commenting on the building's opening: 'the building is an expression in every way of the modern spirit'. In debates about the significance of cities in the development of cultures of consumption, limited attention has been paid to the actual physical fabric of the city (see e.g. Glennie and Thrift 1996; Mort 1996). Consideration of the building has figured more prominently in specific studies of the emergence of the modern department store (e.g. Domosh 1996; Lancaster 1995; Miller 1981) These accounts have highlighted expansions in the size of the building and the use of technological advancements: plate-glass windows and electric lighting to maximize the potential of displays, lifts and escalators to aid circulation, the introduction of steel frames to free large expanses of shop floor from inconvenient partitions. This work pays some attention to the deliberate use of eye-catching buildings, but fails to explore in depth the connections between architectural style and the modes of consumption of fashionable goods. While these cannot simply be read off each other, they are clearly and self-consciously interconnected, and in the design of Simpsons there was a particularly close relationship between architect and fashion entrepreneur. Retail architecture is characterized by a particular ordering of surface, fabric and space which prioritizes the styling of the street frontage and interior spaces. It is through a reading of these two aspects of the architecture that the store's relationship with gender and modernity can be understood.

Novelty was a valuable commodity in the West End, and architectural renewal was an important means by which the area could communicate the fashionable nature of the shopping experience and goods on offer. The earlier part of the century had seen the complete rebuilding of Regent Street, as Nash's elegant Regency stucco gave way to Norman Shaw and Blomfield's much more bombastic imperial parade route (Rappaport 2002). However, the continual piecemeal renewal of the West End, building by building, business by business, could effect an equally significant transformation. Contemporary photographs show London's streets in a state of constant flux during the 1930s: building sites, colourful hoardings and temporary shop buildings were a normal part of the street scene. As part of this process of renewal, Sir James Pennethorne's solid Italianate Geological Museum of 1851 was torn down in 1936, making way for a beacon of international modernism.

Was modernism simply the latest expression of department store novelty? The Simpson Piccadilly building was certainly novel, and pushed back stylistic boundaries in the West End by the use of modernism. The building was extensively discussed in the architectural press, and drew its architectural references from America and Europe, not least the German chain of Schocken department stores designed by Eric Mendelsohn. The store had impressive modernist credentials, with many leading artists and designers collaborating on the project in an unprecedented manner: architect Joseph Emberton; graphic designer Ashley Havinden; former Bauhaus artist László Moholy-Nagy, given charge of the display department; Duncan Miller and Marion Dorn, cutting-edge designers responsible for the carpets. Furniture was purchased from the avant-garde manufacturer Pel. There was even some discussion of using sculpture by Eric Gill and Henry Moore. Before it even opened its doors, the Design and Industries Association was planning to visit, sealing the store's position as an icon of British modernism. In a letter to Emberton, Simpson wittily drew an analogy between the store and an ocean liner, that great archetype of inter-war modernism: 'I feel very much like the captain of a big new ship waiting for the pilot in charge of the tugs to get him out of the dock, so that I can sail spick and span on my maiden voyage' (Simpson–Emberton correspondence, 20 April 1936).

However, it is unlikely that the company's prime motivation in commissioning Simpsons was the furthering of the cause of modern architecture, design and art. The store papers show Alexander Simpson to be a pragmatist in his general approach to retailing. When photographs of the new building appeared in the architectural press, the store's interior was shown as a collection of empty spaces, carefully styled with the 'modern' in mind, furnished with elegant bent ply and tubular steel, fitted with the most up-to-date fixtures, for which Europe had been scoured. However, Simpsons was not just a fashionable modern building bought off the peg; it was carefully designed around the perceived needs of the modern man. These images of the empty store, a conventional form for architectural photography, also reveal a set of stages for the display of garments and for the performances of selling and shopping, routes to draw the customer through the shop floor and send him spiralling up the building's core. Correspondence between store owner and architect show it was planned meticulously to house specific retail practices and precise numbers of garments. One letter from Simpson to his architect (29 August 1935) specifying accommodation for items including 12,000 lounge suits, 300 riding coats, 2,250 hats and 50 suit lengths for bespoke tailoring. Indeed Simpsons was almost literally tailor-made to fit its customer: the design process revealed an almost obsessive preoccupation with the varying dimensions of the male body. Simpson explained one unrealized scheme in the following terms: 'one would arrange to have the wardrobes for, shall we say, regular height men on the ground floor and for the short and tall and stout men on the first floor, and it would not matter whether a man wanted a lounge suit, a sports suit, an overcoat or a pair of pyjamas, we would be enabled in this manner and with the aid of our wardrobe to shew him almost everything we stocked in selling him the one article' (Simpson–Emberton correspondence, undated, probably 1935).

Just as the building's interior spaces were moulded around the Simpsons man and his clothes, the store's façade was also designed to promote consumption. Modernism in 1930s retail architecture was complex and fractured, variations in form and style reflecting differences of use and context. The newly rebuilt Peter Jones department store in Sloane Square was also fêted for its modernism, particularly its ground-breaking structure. The external component of this was the skin of curtain walling, but the owners, the John Lewis Partnership, were known to be less concerned with modern styling than with creating an efficient and flexible retail environment. Simpsons' architectural reputation, on the other hand, was primarily due to its aesthetic innovations. The building was certainly structurally advanced: developments in steel technology allowed an unsupported span of 60 feet across the entrance, reputedly the greatest in London at that time. But the design was famous for the simplicity and horizontality of line on its façade, despite the use of traditional Portland stone veneer. The emphasis on the façade did not weaken Simpsons' brand of modernism or render it superficial. A store's façade was the most crucial part of its design: an advertisement of the store's identity, an instantly recognizable signpost, a focal point for the interface with the street. The façade at Simpsons was used as a canvas for communicating with the street by day and by night, loaded with symbolism about the modernity of the goods and lifestyle that could be bought inside.

Lighting was as important in the vocabulary of modernism as form. Costly experimental work was carried out in the pursuit of suitably show-stopping neon and floodlighting for the Simpsons' façade (see figure 8.2). The result was a façade with bands of floodlighting which could be changed in colour, and a neon sign in Ashley Havinden's specially designed typeface spelling out the store's name, which could be moved around the façade. The frontage was also designed to be decorated: strewn with flags for victory parades, covered in sparkle for Christmas, coated in flowers for a coronation. A modernism which prioritized display, styling and drama is consistent with Emberton's other work, which included buildings at Blackpool Pleasure Beach and the Olympia Empire Exhibition Hall.

Glazing was also a key element in the modernist design canon, and Simpsons featured the classic modernist bands of horizontal glazing across its façade. The innovative concave ground floor window broke on the day after opening, and it proved a very difficult task to find a company willing to insure it. But the glazing had a very important practical function: the large

Figure 8.2.   The Simpsons building communicated with the street at night using cutting-edge lighting techniques
Source: Image of Simpsons in the late 1940s from the Daks archive.

glazed entrance enabled passers-by to see right into the store, and provided an uninterrupted route from the street. The mannequins in the windows were only inches from the pavement, and the special glass prevented reflections, further reducing the boundary between inside and outside. The window display was carefully controlled: not for Simpsons the visual clutter and chaos of so many shop windows. Architect Emberton and head of display Moholy-Nagy specified a range of simple modern display fittings, changeable window backs and bases in a limited palette of plain colours and natural wood, and also that text was to be confined to a small glass noticeboard.

If the brand of modernism employed was closely linked with the selling of clothes, it also reflected the building's immediate locality. The Crown Commissioners stipulated the traditional Portland stone of the façade and the signage was closely negotiated with the London County Council. Emberton's buildings for His Master's Voice in Oxford Street used a more exciting range of modern materials, such as shiny Vitrolite and glass brick. Nonetheless, Simpsons' façade made it stand out from the flanking buildings on Piccadilly. Its boldness helped align Simpsons with the flagship fashion stores of the West End, and assured its place on their route. The rear elevation also drew attention to the difference between Simpsons and its Jermyn Street neighbours, identifying it as the home of a distinctively modern masculinity. However, the quality of materials and luxuriously styled elegance of exterior and interior retained something of Jermyn Street's exclusivity. This double-sidedness helped to shape the identity of the 'Simpsons man' in an age that was witnessing the increased democratization of fashion.

The relationship between architecture and clothing in the 1930s was complex, coloured by the still influential writings of modernists such as Loos and Le Corbusier denigrating women's fashions in particularly harsh terms (Breward 2001; Wigley 1995). Yet the success of Simpsons shows that a certain type of modernist style could sit very happily with fashion, if that of men. This perceptible reassessment of retail architecture, formerly considered somewhat disposable, ephemeral, and rather beneath the attention of the architectural establishment, was not unrelated to the emergence of new masculine consumer identities.

The Simpsons Man

It was no accident that the store was opened by a popular British sporting hero. Simpsons both drew upon and helped construct a new kind of metropolitan masculine consumption. The stock suggested a wardrobe for a particular kind of English lifestyle: lounge suits, dinner suits, full dress suits, formal morning suits, tropical suits, and sports clothing, including plus fours, golf suits, and riding clothes, which could be tried on with the aid of a dummy horse in the changing room. The clothes on the racks were infused with images of the Simpsons man, as he appeared in Ashley Havinden's store advertisements and in Moholy-Nagy's shop window displays. This ideal Simpsons man was modern, urban and sporting. He was from the wealthy middle classes, most likely working in the city, but with country pursuits. He was not a dandy, but was always well dressed, and confident of being so, having chosen the clothes himself. He had square-jawed, suntanned good looks and was self-assured with beautiful women, often to be found impressing one with his 'Daks appeal' (as it was described in a 1938 advertising campaign). But he was most at home in the company of other men, in a sports-jacketed crowd at the races or chatting and smoking in a dinner suit at his club. His clothes showed him to be most definitely English, but with a hint of more relaxed American style. He was equipped to travel the world, but, crucially, his clothes were bought in the West End of London. The specific location of the store permeated the fabric of the garments, permanently branded with the name of the street, 'Simpson Piccadilly'.

However, the relaxed, confident manner of this modern West End habi-tué belied the fact that he provoked anxieties about class and gender, tensions which were housed in his Piccadilly home. Alexander Simpson's obituary in the *Sunday Express* (16 May 1937) described his goals: 'He had visions of conquering the West End trade in men's wear. He dreamed of owning a store where men would be able to obtain things that are normally only sought in the most exclusive and most expensive West End establish-ments at competitive prices.' Although the store offered bespoke tailoring, stress was laid on the quality, variety, cut and range of fittings of the ready-made clothing which made up the bulk of business, creating a more affordable 'Savile Row' look. For the project to succeed, Simpsons needed to maintain a delicate balance between the cachet of the store and the large volume of sales required to justify the store's scale. This was a new dilemma for menswear retailing, but a familiar issue for the department store (Lancaster 1995; Miller 1981).

The extra business was sought from those who aspired to be the Simpsons man, but who had previously been excluded from his wardrobe. Suggestions were mooted by the Board (Minutes, 17 February 1937) that information on the correct way to dress should be incorporated into store catalogues. A series of early store advertisements played on the insecurities of a series of men in socially aspirational positions, including solicitors and architects. One advertisement in the *Evening News* (21 June 1938) featured a man in an ill-fitting tweed jacket and baggy trousers courting a chic woman with a decidedly dissatisfied expression on her face: 'A fine romance . . . yes . . . but what a pity he doesn't get his clothes at . . . Simpson, Piccadilly' (ellipses in original). Fears of acceptance into traditional English elite circles were also played upon in advertisements directed at tourists and expatriates: 'however stoutly the Englishman may affect a careless indiffer-ence towards dress . . . some of the worst predicaments in English life are associated with "the wrong clothes" . . . The unforgivable sin is not doing the wrong thing, but doing it in the wrong clothes' (*Ocean Times*, July 1936). Alexander Simpson's mission was to ensure that the store itself was correctly dressed for its geographical location, and he wrote anxiously to the architect during the planning stages: 'With regard to your suggestion for the name on the Jermyn St front, do you think it is wise to have Neon in this street?... It seems to be against the character of a high-class shop in Jermyn Street' (31 March 1936).

The *Manchester Guardian* was sceptical that the store could attract sufficient custom for another reason: shopping was still believed to be a substantially female preserve.

> Men, so rumour has it, dislike shopping. Even the most stout-hearted fellow flinches when it comes to entering a department store, while the majority of them entreat their wives, their mothers or their sisters to buy their gloves, their shirts, or collars for them. If that be so, is not the opening of a large store for men... foredoomed to failure? (1 May 1936)

The new department store format adopted by Alexander Simpson was on a quite different scale from earlier West End men's outfitters and tailors. It introduced very different 'feminized' consumption practices involving window-shopping, browsing and impulse buying. *The Lady* (7 May 1936) commented on this:

> It is amusing to find that the man's shop is designed and set out with all the allure of one devoted to women's luxuries. Shopkeepers, evidently, do not share that masculine theory that a man always knows just what he wants and so is immune from display or advertisement.

Coverage in the press showed that the store clearly provoked anxieties about 'manly' behaviour, frequently hinting at the emasculation of the new male shopper. Simpsons addressed these concerns directly, tirelessly promoting this new kind of men's shopping. Abstract modern window displays were very different from anything to be found in Jermyn Street, but were much more in line with Regent Street and Oxford Street stores. Once inside the store displays were designed to tempt the customer into making several purchases. Alexander Simpson stipulated to Emberton (29 April 1935), items for impulse buying should be placed near the main clothing displays: 'for example next to the wardrobe containing blue suits we would have say half a dozen shirts; a dozen ties; 3 pairs of socks, and a couple of hats, and a few handkerchiefs; and so on between each wardrobe...'.

Simpsons was pioneering shopping as an acceptable and pleasurable leisure pursuit for men. The *Daily Mail* (30 April 1936) described the environment as 'nine-floors of store exclusively for men, where they can revel in all the delights of shopping'. But a trip to Simpsons was about more than just shopping, the snack bar and club room providing the male equivalent of the tea rooms and restaurants of Harrods and Selfridges. A series of early store advertisements entitled 'Sorry I'm late – but look what I got at Simpsons!' explicitly used the previously feminine language of desire, temptation and loss of oneself in the retail environment, as in this example from 1936:

> When a man tells you he's 'just popping into Simpsons for a stud' don't expect to see him come popping out a minute later. For once a man is in Simpsons he is apt to lose all sense of time. Blame the wonderful barber's shop if you like for tempting him to a shave. Blame half the snack bar for having half his friends in it. Blame the aviation exhibition on the fifth floor for showing him a flying flee. Blame our shirts, our ties, our shoes – for bewildering him with choice. But don't blame him: it's not fair. After all, what man can tear himself away from Simpsons? (*Evening Standard*, 6 May 1936)

Another advertisement in the series (*The Tatler*, 12 August 1936) addressed a female audience, 'When husbands buy anything at Simpsons – don't think that that's just the end of the matter. Good gracious, no! You've got to *admire* the purchase. You've got to discuss it with him – in all the glowing terms that it deserves. He expects you to! . . . '. The concerns about threats to heterosexual masculinity were sent up and neutered, the enthusiastic shopper pictured rushing safely back to the his wife's bed in his new Simpsons dressing gown (figure 8.3).

The gendering of space was the key issue in seeking to legitimize this new shopping masculinity, as expressed in the *Manchester Guardian*: 'the reason why men fight shy of shopping is that they feel that they are entering alien territory, from which they may not only be expressly excluded but to which they are only admitted on sufferance' (1 May 1936). The store needed to be a recognizably masculine space, distinct from the other West End department stores. Simpsons claimed that it had created a 'home' for men, a 'man's world' within the West End. To achieve this the store used its location, its brand of modernism, its chic but comfortable interiors furnished with rugs and armchairs, even its system of selling, designed to replicate a gentleman selecting garments from his own wardrobe: 'If I use my wardrobe at home, I have everything in it . . . my town suits, my country suits, my three or four overcoats, dressing gown, bedroom slippers – in fact everything that I need, and I believe that it is the best policy to work in this way' (Simpson–Emberton correspondence, 9 July 1935).

But a men's enclave in this location ran a high risk of infiltration, and despite the best efforts of the store's management, it remained a space where masculinity and femininity were in tension. The one concession to femininity on opening day was the gift shop where women could purchase items for husbands, sweethearts, sons and fathers. But in the summer of

Figure 8.3. The anxieties raised by modern masculine consumption practices were addressed in a series of advertisements in the late 1930s
Source: Daks archive.

1937, barely a year after opening, a women's department opened in the 'man's world', soon spreading over two floors of the building and hailed by the Board as 'quite possibly the salvation of the store' (Minutes, 26 October 1939). Changes in the gender balance of London's shopping population in wartime may have had some influence on the success of this development, but it was reported that women were shopping in significant numbers at Simpsons from its opening in 1936. It was in the nature of the department store that the arrangement of the constituent departments was unstable, but the allocation of floor space to the women's department was always hotly contested. Despite good business, the company was half-hearted in promoting the women's department and was keen to 'confine' it to one floor only as soon as the war was over. The department was seen as a threat to the store's image, the Board of Directors expressing concerns that 'the women's floor must have a distinct character, but it must not deviate from the general character of the store and must be in keeping with the store and arise out of the store' (Minutes, 24 November 1938). The fear was that allowing women access would upset the meanings of shopping at Simpsons, which depended on it being a distinctively masculine space. The gender-informed remapping of Simpsons' departments also suggests that the management had made a series of miscalculations about masculine consumption practices. Breward's work (1999) on the nineteenth-century male consumer has established his agency in the creation of complex, fashion-conscious identities which required sophisticated choosing processes. While such a group undoubtedly still existed in the 1930s, Simpsons' promotional campaigns suggest there was anxiety about there being enough confident and sophisticated male consumers to fill the store. Alexander Simpson instigated a re-education programme combining the design of the retail environment with innovative selling methods to bring the cavernous spaces to life – and to recover his considerable investment.

Masculinity and Metropolitan Modernity at Simpson Piccadilly

The opening of Simpson Piccadilly was a specific moment when architectural modernism, modern masculinity and modernity converged in the West End of London. The unstable relationships between place, design, retailing and gender mean that the changing geographies of West End can be used as a way of understanding shifting notions of modern metropolitan masculinity. The store adopted a version of modernity which allowed for experimentation in gendered retail practice, which was able to encompass tensions between exclusivity and democracy, and which found a visual expression in a stylish form of modernism. This was above all coloured by its West End location. That this melting pot should be at 202 Piccadilly was crucial, as it allowed for dialogue with past consumption patterns, thanks to its position on an intersection between old and new shopping routes, between the homes of traditional and modern fashionable English masculinity.

But fashion retail requires novelty and reinvention, and from the 1950s the location of fashionable modern masculinity began to move to other West End sites, such as Carnaby Street. Simpsons' relationship with modernity was renegotiated and the store's reputation graduated towards the provision of good-quality traditional men's clothing. The company did not attempt to give the store an architectural facelift, instead cherishing what had become an iconic building of a bygone age, a symbol of classic good design. As the decades went by, this resulted in a perceptible reorientation of business towards the traditional Jermyn Street model, culminating in the closure of the Piccadilly store in the late 1990s, and the retention of the later Jermyn Street extension as a smaller Daks shop. The convergence of modernism, modern masculinity and modernity at Simpsons had proved fleeting, but the building remains as a monument to the 'modern spirit' which was created there in 1936.

NOTE

1   Historical material relating to Simpson Piccadilly referred to in the text is from the Daks Archive, unless otherwise stated. The assistance of the archive in the conduct of this research is very gratefully acknowledged.

REFERENCES

Breward, C. 1999: *The Hidden Consumer: Masculinities, Fashion and City Life, 1860–1914*. Manchester and New York: Manchester University Press.

Breward, C. 2001: Fashioning masculinity: men's footwear and modernity. In S. Benstock and S. Ferris (eds), *Footnotes: On Shoes*. New Brunswick, NJ and London: Rutgers University Press, 116–34.

Domosh, M. 1996: The feminised retail landscape: gender ideology and consumer culture in nineteenth-century New York City. In N. Wrigley and M. Lowe (eds), *Retailing, Consumption and Capital: Towards the New Retail Geography*. Harlow: Longman, 257–70.

Glennie, P., and N. Thrift 1996: Consumption, shopping and gender. In N. Wrigley and M. Lowe (eds), *Retailing, Consumption and Capital: Towards the New Retail Geography*. Harlow: Longman, 221–37.

Lancaster, W. 1995 *The Department Store: A Social History*. London and New York: Leicester University Press.

Miller, M. 1981: The bon marché: bourgeois culture and the department store, 1869–1920. Princeton: Princeton University Press.

Mort, F. 1996: *Cultures of Consumption: Masculinities and Social Space in Late Twentieth-Century Britain*. London: Routledge.

Nead, L. 2000: *Victorian Babylon: People, Streets and Images in Nineteenth-Century London*. New Haven and London: Yale University Press, 2000.

Rappaport, E. 2000: *Shopping for Pleasure: Women and the Making of London's West End*, Princeton: Princeton University Press.

Rappaport, E. 2002: Art, commerce, or empire? The rebuilding of Regent Street, 1880–1927. *History Workshop Journal*, 53, 94–117.

Wigley, M. 1995: *White Walls, Designer Dresses: The Fashioning of Modern Architecture*. London and Cambridge, MA: MIT Press.

Chapter Nine

Mosques, Temples and Gurdwaras: New Sites of Religion in Twentieth-Century Britain

Simon Naylor and James R. Ryan

In August 1995 the Swaminarayan Hindu Mandir, a massive, glistening building of white marble and sandstone with six domes and intricately carved pillars, was officially unveiled in Neasden, north-west London (figure 9.1). The construction and opening of this traditional-style Hindu temple (or *mandir*) was surrounded with unprecedented publicity. The building, dubbed 'Neasden's Taj Mahal' in the national press (*The Independent*, 17 August 1995, p. 12), won several environmental and architectural awards and was widely recognized as one of the most impressive examples of religious architecture to have been constructed in Britain in the second half of the twentieth century. The fact that this beautiful and expensive landmark was a Hindu sacred space rather than a Christian building was widely noted. Indeed, the last Christian buildings of comparable stature, the Anglican Cathedral in Coventry and the Catholic Metropolitan Cathedral in Liverpool, were completed in the 1960s. The opening of the Swaminarayan Hindu Mandir made a dramatic intervention not only in the visual landscape of suburban London, but in the discourses of home, community and belonging that shape such landscapes. The publicity surrounding the opening – generated in part by the Swaminarayan Hindu Mission itself – drew the site to the attention of local and national politicians, royalty and the global media. Together with the large Haveli Cultural Complex and visitor facilities to which it is attached, the mandir attracts thousands of visitors each year. Indeed, the Swaminarayan Hindu Mandir in Neasden is perhaps the most emblematic statement of the vibrancy and permanency of South Asian culture, religion and identity within Britain in the late twentieth century. It brought to the attention of a wider British public the remarkable developments of South Asian reli-

Figure 9.1.   The Swaminarayan Hindu Mandir in Neasden, north-west London, opened in 1995

gious architecture, which has transformed the built environment of many British cities and suburbs. Yet behind the headlines generated by the opening of such spectacular edifices lies a much longer history and geography of non-Christian religious groups in Britain. Indeed, the impact of such groups on the British landscape may be traced back several centuries. However, the most dramatic transformations in the ethnic and religious make-up of Britain have occurred in the twentieth century.

This chapter charts these new landscapes of religion and ethnicity as they developed in Britain in the twentieth century. Religion has played a central role in the construction of twentieth-century Britain, whether we wish to consider its role in the affairs of the state, in the case of the Church of England, or in the everyday life of the country's inhabitants. Even within a society that is perceived as becoming increasingly secular, religious values, practices and beliefs shape many aspects of social life, from language to law (see Holloway and Vallins 2002). It is thus surprising just how little attention historical geographers have paid to the place of religion in the making of modern Britain. For many it seems that the common assumption that Britain is becoming an increasingly secular society holds sway. One recent volume of historical geography essays has three references to Christianity in the index but none for 'religion' (Graham and Nash 2000). This appears to echo the fact that many historical accounts of religion in Britain are concerned exclusively with Christianity (Gilbert 1980; Hastings 1991). To be sure, Christianity has been a central cultural force in Britain for many centuries. Yet the conventional idea of Britain as an exclusively Christian place was challenged dramatically in the twentieth century. By the end of the twentieth century sizeable minority populations of Muslims, Hindus, Sikhs, Jews and other non-Christian groups existed in Britain. This chapter considers the evolution of places of worship for Muslims, Hindus and Sikhs. By the 1990s these faith communities were, after Christians, the three largest religious groupings in Britain (Weller 1997: 30).

The cultural and religious make-up of Britain has been transformed through a range of complex forces, including colonialism, immigration, trade and cultural exchange, that operate on a number of temporal and geographical scales. It would be a very narrow historical geography indeed that did not explore how such currents of modernity have linked Britain to other places. Even early twentieth-century geographical accounts of Britain, such as Halford Mackinder's *Britain and the British Seas* (1902) included accounts of 'imperial Britain' and 'new Britains across the ocean' (1902: 345). Indeed Mackinder's whole conception of Britain was as an imperial nation lying at the hub of 'a vast imperial nodality' (1902: 356). While historical geographers in the twenty-first century reject Mackinder's vision of imperial unity, they remain centrally concerned with the temporal and spatial interconnections and relationships between places. Indeed, it is not possible to read the multicultural map of modern Britain without appreciating the complex and long-standing connections between places and people within Britain and the rest of the world.

This chapter starts from the premise that a central and fundamental part of any historical geography of twentieth-century Britain should be the mapping of the nation as a multicultural, multiracial and multifaith society. However, despite a wealth of work on British imperialism, colonialism and modernity by historical geographers in recent years, little attention has been paid to ethno-religious minorities in Britain. In part this may be because such issues are seen as being the domain of social and cultural geographers or those working in the 'geography of religion'. This latter field witnessed a significant resurgence of interest in the 1990s, spawning important studies of sacred space and the performance of religious identities (Kong 1990, 1993, 2001a, 2001b; Raivo 1997a, 1997b; Park 1994; Yeoh and Hui 1995). Associated work by cultural geographers has begun to explore important questions of identity amongst ethno-religous minority groups (Dwyer 1996, 1999; Vertovec and Rogers 1998). However, work on the history of religion in twentieth-century Britain that considers religions besides Christianity has been left largely to historians of religion, scholars of religious studies and historically minded sociologists (see e.g. Badham 1989; Eade 1996; Parsons 1993; Thomas 1988; Wolffe 1993, 1994; Knott

1986). In this chapter we seek to redress this balance by showing how Britain's ethno-religious landscapes were transformed in the twentieth century by powerful currents of modernity that forged ever more complex interconnections between places and between peoples.

Muslim, Sikh and Hindu Places of Worship in Britain

One way to chart the profound changes in the geography of religion and ethnicity in Britain is through the growth of non-Christian places of worship. Religious buildings are significant as spaces of religious practice as well as sites of symbolic meaning and cultural identity. Religious buildings, as Lily Kong has noted, 'may be invested with symbolic meanings – personal, sacred, social, and political' and deserve greater attention by scholars of religion and cultural geography (1993: 23). The geography of religious architecture in Britain changed dramatically during the second half of the twentieth century when the number of Muslim mosques, Hindu mandirs and Sikh gurdwaras increased steadily, accelerating in the last two decades of the twentieth century. Religious buildings not only embody the traditions and meanings of a particular religious faith but also command a presence for it in the local, regional and national landscape. Tracing their history and geography is therefore potentially a very useful way of charting growth in the number, prosperity and confidence of minority religions. It is also a means of assessing the extent of organized religious activity. However, there are both conceptual difficulties and practical problems of sources that underlie this work. For instance, though we focus here on the three largest minority religious groups we would not wish to underplay the diversity of religious affiliation in modern Britain. Indeed, late twentieth-century Britain was characterized by the dramatic growth of a range of 'new' religions and spiritual movements (Holloway 2000). Nor would we wish to conflate specifically religious affiliation with ethnic identity. The boundaries between religion and ethnicity are both dynamic and complex, and not necessarily coterminous. We move on now to trace in more detail the historical geographies of sites of religious minority worship in Britain.

The Spatial Evolution of Mosques, Gurdwaras and Mandirs in Britain

Early twentieth-century Britain was not a uniformly Christian landscape. Indeed it was home to a small and growing number of non-Christian religious communities. Moreover, such a presence was by no means new. Since the Middle Ages currents of trade, commerce and, in the eighteenth century, slavery had brought individuals and groups of non-Christians, particularly Muslims and Jews, to settle in Britain (Matar 1998). By the nineteenth century, as historians have shown, Islam and South Asian religions had a significant (if numerically small) presence in Britain, notably in major trading cities such as Bristol, Liverpool and London (Beckerlegge 1997). The latter, as the great imperial metropolis, was particularly significant as a space for the development and interaction of different ethnic and religious communities (see e.g. Burton 1998). Indeed, questions of ethnicity, race and immigration, which came to have such significant bearings on the twentieth century, had strong roots in the previous century.

The early history of non-Christian faith groups needs also to be set within the backdrop of religious debate, notably between different denominations and factions within Christianity. Historians have thus charted how the nineteenth century witnessed a decline in the power of the established church and the social and legal recognition of other faith groups, including Roman Catholics, non-conformists, and Jews, as well as secularists (Beckerlegge 1997). Removal of religious qualifications for entry into Oxford and Cambridge in 1871 helped foster a growing presence of students from the Indian subcontinent, most famously perhaps Mohandas Karamchand Gandhi. It was these flows of empire, of people and knowledge that helped establish some of the early non-Christian places of worship in Britain.

The earliest of these buildings were mosques, such as the Liverpool Mosque and Muslim Institute (1887) and the Shah Jehan Mosque (known also as the Woking Mosque) in Surrey (1889). The Liverpool Mosque was the brainchild of Henry Quilliam, a convert to Islam following his travels in North Africa, whose cloistered mosques the Liverpool Mosque was designed to echo. Similarly, the founder of the Woking Mosque was Dr Gottlieb Wilhelm Leitner, a Hungarian orientalist who wanted to attract students to his Oriental Institute established in Woking in 1884. The Liverpool Mosque flourished until 1908 when Quilliam left Britain. It subsequently closed. The Woking Mosque closed for similar reasons in 1899 but was reopened in 1912 by a Muslim mission from the Punjab.

The revival of this particular building signals the slow beginnings of non-Christian places of worship established and maintained by South Asian religious communities, rather than European or British converts and orientalists. Thus the Fazl Mosque – often referred to as the London Mosque – in Southfields, established in 1924 and the oldest mosque in London, derived from the Ahmadiyyas Movement (founded 1889) sending a missionary to Britain in 1914. This example indicates how Britain's imperial connections shaped its own religious landscapes. The forging of this Muslim community and the building of this mosque blurred conventional boundaries between core and periphery, modern and traditional. At the time of its opening, the mosque was regarded in the local and national press as both highly exotic and, simultaneously, as modern. It was exotic because Islamic religious and architectural traditions were commonly associated with 'the Orient' and the exotic landscapes of empire (Crinson 1996). However, it was also modern in its innovative concrete and steel construction and its architectural design, and compatible with its setting in suburban London (for a detailed discussion see Naylor and Ryan 2002).

The early twentieth century may also be seen to mark a shift towards more stringent attitudes to immigration. Victorian Britain had, by contrast, been more open to immigration, if only because so many of Britain's population were themselves migrating to make 'new Britains' overseas. Increasing restrictions were placed on immigration, most notably through the 1905 and 1919 Aliens Acts. However, the years following the Second World War witnessed a return to mass immigration to Britain, fostered mainly by post-war labour shortages. While it is possible to locate in these years the birth of a new 'multicultural' Britain, the actual numbers involved were still relatively small. Indeed, the ethnic minority population in 1955 was approximately 100,000, and few foresaw (either positively or negatively) the landscape of multicultural, multiracial and multifaith Britain as it was to develop over the next half-century.

Immigration into Britain of South Asians, from whom much non-Christian religious tradition in Britain developed, grew strongly from 1962 up to the Commonwealth Immigrants Act of 1968. It is perhaps unsurprising that the steady growth of mosques, temples and gurdwaras from the 1960s corresponds to this phase of immigration. Places of worship were established in the main centres of South Asian settlement in London, the Midlands and the north of England. The first officially listed Sikh gurdwaras were in Liverpool and Newcastle, in May 1958. Others followed in east London (1960), Manchester (1961), Birmingham (1962), Doncaster (1962), Leeds (1963), Nottingham (1963), Gravesend (1963), and Southall (1964). Similarly, the first Hindu mandir to be officially registered was based in Leicester in 1969, although both Hindus and Sikhs had been making use of their own, non-registered, places of worship for several decades prior to this. By 1980 the numbers of officially registered mosques, gurdwaras and mandirs in England and Wales had grown from 12 in 1964 to some 232. However, at this stage many non-Christian places of worship were not new buildings but conversions of existing buildings, including those formerly used by Christian churches. (Planning permission was not required to convert a religious building from one faith to another.) By the year 2000 the number was almost 1,000, with the majority having been established during the previous 20 years. These last two decades of the twentieth century thus witnessed the establishment of the greatest number of mosques, mandirs and gurdwaras. By far the greatest growth was in the number of mosques, which grew from 9 in 1965 to 296 in 1985 and to 614 in 1998 (see figure 9.2)

Much of this expansion occurred within established communities and reflects their growing economic and social position in particular localities. In particular, many of the new sites of worship were the result of processes of sectarianism, where communities of practitioners became numerous and wealthy enough to sustain traditional religious preferences. For instance, Sikh men settling in Southall, west London, in the 1950s shared a single gurdwara. However, as more Sikh communities grew and established themselves in the area new gurdwaras emerged to serve particular ethnic and cultural affiliations. This is also the case amongst the Muslim population, with Sunni mosques often dividing into Barelwi or Deobandi, for instance, which connote different interpretations of Islamic scripture.

This evolution and expansion of non-Christian religious features in the British landscape was highlighted by the decline in Christian worship throughout the twentieth century. While the 'secularization thesis' has been the subject of considerable debate, it certainly seems the case that the heyday of Christian church attendance and church-building was over

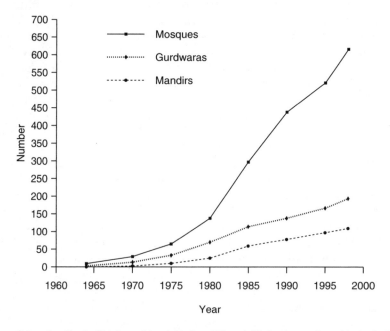

Figure 9.2. Growth of officially registered Muslim, Sikh and Hindu places of worship in England and Wales, 1964–1998 *Sources: Marriage, Divorce and Adoption Statistics,* ONS 1964–98; 'Ethnicity and Cultural Landscapes' project survey, School of Geography and the Environment, University of Oxford, 2000.

by 1900 (although see Brown 1996). The common trend in the twentieth century was of declining church attendance and, more importantly, of the increased perception that decline in attendance was accompanied by decline of traditional 'Christian values'. By the 1960s it was clear to many that Britain was no longer just a Christian country, a fact that many groups were far from comfortable about recognizing. Ironically, the growing sense of Britain as a multicultural and multifaith nation, where non-Christian places of worship were not out of place, was thus tempered by very real difficulties and conflicts around the idea and practice of multiculturalism. As the Salman Rushdie affair in the mid-1980s showed, considerable conflict remains between liberal individualism and ethnic tradition, between freedom of expression and freedom of ethnic minority religious sensitivity, and between 'the idea of Christian Britain with its established church and the concept of a religiously plural society' (Kushner 1994: 420).

Sites and Scales of Minority Religious Worship

At regional and city scales, distribution of the sites of worship for a particular religion inevitably reflects the distribution of that religion's wider ethnic community. Within regions and cities the geography of places of worship again reflects historic patterns of settlement as well as the economics of housing and building. The location within particular cities is determined largely by the location of ethnic minority communities as well as pragmatism in terms of where land, building and planning permission have been available. Thus, for instance, the concentration of Hindu mandirs in Greater London can be explained by the fact that London has the largest Hindu communities. Nonetheless, early official records show just how mobile places of worship have been, particularly in the early years of a community's existence. Many Islamic places of worship in the 1950s and 1960s moved frequently, sometimes after only a matter of weeks. They might move, for example, from a basement of one building to two rooms at another address. Communities also moved between different cities. The 'Zaoula Islamica Allaoula', for example, first registered as a mosque in Hull in 1944, and moved a further three times between 1959 and 1967.

There is a micro-geography that might therefore be traced, that operates at the level of streets and even buildings. Research on the construction of Muslim, Sikh and Hindu places of worship in Britain shows them to be of tremendous variety and diversity. In building types and styles alone places of worship range widely, from adapted private dwellings to former Christian sites of worship, from converted cinemas, warehouses and offices, to railway arches, squash courts, and even old military arsenals. This diversity in building form seems be structured along a general evolutionary pattern, whereby particular religious communities, as they have developed in size as well as in social and economic security, have moved from adapted private dwellings to the conversion of larger vacant public buildings. The 1980s and 1990s witnessed the erection of grand purpose-built places of worship, notably in cities such as Birmingham, Derby, London, Leicester, Blackburn, Leeds and Bradford, the Neasden Mandir being just one of many such sites.

While these specially commissioned buildings may be spectacular, they are not typical. Indeed, the majority of mosques, gurdwaras and mandirs in Britain are housed within adapted buildings. The external architectural appearance of these buildings thus frequently runs counter to their content, and many places of worship remain inconspicuous to those outside the community that use them. The interior spaces of these buildings frequently undergo a complete metamorphosis: former squash courts, cinemas and social clubs are transformed into sacred spaces. Thus a former Irish club in Chatham is now a mosque and a former Christian church in Southampton is now a gurdwara. Through this long-standing process individual buildings acquire complex biographies as sacred spaces. Perhaps the most famous example is the Jamia Masjid (Mosque) in Brick Lane, east London, housed in a converted synagogue, itself converted from an original Huguenot church building (see Eade 1996).

Similar diversity also exists in these sites' local morphology and setting. While most are found in inner-city areas where ethnic minority populations are often concentrated, increasing numbers are found in suburban residential areas and on light industrial or commercial sites. Explanation for any such discernible patterns that reside in these distributions lies in the particular and often complex circumstances surrounding the purchase of suitable property and the granting of planning permission as well as in the particular histories of the communities in question (Gale and Naylor 2002).

Contested Landscapes

The historical geography of religion and ethnicity in twentieth-century Britain is tied closely to the politics of place (cf. Amin and Thrift 2002). Indeed, the establishment of places of worship for ethnic minorities has invariably been achieved in a climate of conflict and contestation (Dwyer 1996). A number of reasons exist for such conflicts. To begin with, as a number of commentators have observed, English law has not effectively safeguarded the freedoms of religious minorities to observe their faith without restrictions and intolerance, although the situation may change with the incorporation of the European Convention on Human Rights (cf. Nye 1993).

Another spur to conflict comes where local residents mobilize opposition to planned developments. The Swaminarayan Hindu community, for example, faced decades of opposition in its attempt to construct places of worship in suburban London. Before the building of the Swaminarayan Mandir in Neasden the temple community went through a long, drawn-out and ultimately unsuccessful struggle to build a mandir at a site in Harrow, close to the majority of temple users. The construction of the mandir in Neasden – relatively far from the originally intended site – has also been marked by controversy and conflict. Local residents and resident groups developed new senses of territoriality in response to what they perceived as a threat to their homes, public areas and community. For a number of local residents, the mandir was a visual sign of intrusion into and invasion of a predominantly white British space (Rogoff 1998). Local community organizations, notably Brentfield Tenants' and Residents' Association, co-ordinated local residents' opposition to the development and expansion of the mandir.

Controversies over the design, location and use of such buildings are not new and have long been part and parcel of wider forms of prejudice and discrimination against minority groups in Britain. When the Shah Jehan Mosque in Woking was built in 1889 *The Building News* reported the 'great controversy has ensued because the mosque . . . has a dome . . . as many aver that domes are only admissible over tombs or commemorative monuments' (2 August 1889, p. 142). The Liverpool Mosque, opened in 1887, also faced complaints about the 'public advertisement' and practice of Islam in Liverpool from local publications such as the *Liverpool Review* (Beckerlegge 1997: 253). In 1891 the mosque moved out of the city centre to a property in Brougham terrace, on the outskirts of the city. Following this move, the call to prayer (*azan*) from the building's balcony produced demonstrations by local residents and stones being thrown at the *muezzin* (the person performing the *azan*). Such hostility stemmed not simply from irritation at the timing and volume but rather, as Beckerlegge (1997: 253) has noted, quoting local press reports at the time, 'it served as a reminder of the "Eastern Humbug" which was "detested" by "Western Folk"'. Protests from local residents at the call to prayer in Liverpool in the 1890s are echoed by protests over the daily broadcast of the *azan* from the East London Mosque in the 1980s and the fear of the 'Islamization of space' that often motivates such protest (Eade 1996; Metcalf 1996). There are numerous instances where different groups have contested religious minority places of worship (Dwyer 1996). The London Mosque had its expansion plans thwarted, in part, by the campaigns of local residents, who argued that the site no longer participated in the local community, and that increased numbers would put an unbearable strain on the local area. These arguments were supplemented by a range of derogatory remarks about the mosque and its users in the local press (Naylor and Ryan 2002). The ISKCON Hindu Temple, near Letchmore Heath, north-west of London, experienced similar problems, although it was ultimately more successful at defending its religious needs than the London Mosque (Nye 1993).

Transnational Flows

While places of worship are undoubtedly caught up in local circumstances they are simultaneously traversed by transnational flows of culture, capital and communication. Places of worship may thus usefully be read as a localized manifestation of the globalization of ethnic cultures (Eade 1997). Take again the example of the Swaminarayan Hindu Mandir. As part of the Swaminarayan Hindu Mission (and the wider Hindu community) the mandir has extensive global links. The building was an international collaboration in more ways than just its physical construction. Flows of people, faith and information tie its headquarters in Gujarat in India and to other Mission centres worldwide – from California to Kenya. It also has a virtual existence promoted on the internet (by AksharNet in the USA and the UK). Mission publications stress the mandir's traditional architectural form, archetypal Nagara style of North India based on the Mandala sacred plan, and grand design as a means, through a global network of centres, of uniting a diasporic Hindu community and revitalizing faith and unity.

Moreover, the erection of a Hindu mandir at the heart of London's suburbia represents a dramatic development amidst mundane residential properties, yet one that nonetheless ties the geography of this space to long historical processes of empire and global cultural exchange. The suburbs of London, in particular, are products of the expansion of the imperial metropolis in the late nineteenth century and reflect, in both architecture and urban form, cultures of empire (Driver and Gilbert 1997). As we noted above, two of the earliest mosques in England were established in London's suburbs – the Shah Jehan Mosque in Woking and the Fazl Mosque in Southfields. Global economic shifts in the division of labour and patterns of migration have produced sites of convergence of different ethnicities within the urban and suburban landscapes of major cities, from Vancouver to London. Although such convergence has often resulted in conflicts over differing attitudes towards types of domestic architecture and landscape (cf. Ley 1995), it has also generated new hybrid architectural styles, the Fazl Mosque's exotic-modern design being perhaps the best, but certainly not the only, example of this.

The twentieth century has also witnessed the development of new kinds of spaces for the performance of religious and ethnic identities. The 1995 opening ceremony for the Swaminarayan Hindu Mandir in Neasden began with a parade through central London, involving Swaminarayan Hindu groups from around the world in a performance and celebration of Hindu cultural traditions. The mandir has become a major religious and tourist attraction and provides a means of displaying one kind of religious tradition to new audiences coming from around the world. New kinds of technology, notably television and the internet, also open up new spaces for the communication of religious ideas and performance of ethnic and religious identities (Kong 2001b).

In another sense, however, religious minority groups may be thought to inhabit different worlds within Britain. Ethno-religious minorities suffer from higher levels of unemployment, poorer housing, and a worse standard of education than white ethnic groups. They are also frequently victims of religious and racial prejudice (Modood et al. 1997; Mason 2000). Indeed, the picture at the start of the twenty-first century for British ethnic minorities is not a rosy one. Britain has seen the rise of political support for far-right groups such as the British National Party, a widespread popular hostility to 'asylum seekers', and riots in cities such as Bradford in the summer of 2001. Thus for many British Hindu, Sikh and Muslim communities the new millennium does not herald the end of religious and ethnic prejudice; it merely marks out the terrain for new sets of struggles.

Conclusion

The task of tracing the historical geography of ethnic minority religions in Britain in the twentieth century is a complex one. In this chapter we have focused on aspects of the historical and geographical evolution of Muslim, Sikh and Hindu communities. A broader study would need to take into account the full range of religious minorities, including Jews, Zoroastrians, Baha'is, Afro-Caribbean (especially evangelical) Christians, independent Christian churches, new religious movements (of which there have been some 450 since the Second World War) and New Age 'religions'. It would also need to take into account the range of spaces across which different groups forge collective geographies of faith and identity, including the growing communication networks within and between different religions, such as the various inter-faith movements (see Weller 1992, 1997).

Of course, the task of measuring religious practice and ethnic affiliation is fraught with conceptual difficulties. Official national statistics and census data provide only a very partial and incomplete means of ascertaining the size of religious minorities and degree of active participation in religious faith and institutions. In Northern Ireland, for example, figures for identification with a religious denomination have long remained high (90 per cent in 1971). However, even these measures are problematic. The 2001 census, for example, asked residents of Northern Ireland for their religious identification only in terms of Catholic or Protestant. Unlike other parts of the UK, non-Christian religions were not identified. Moreover, legislation on religious freedom and racial discrimination has been applied unevenly across the UK. Unlike England and Wales, Northern Ireland has had legal protection against incitement of religious hatred, but did not have race relations legislation until 1997.

Official governmental registers of places of worship underestimate their total number, due to the voluntary nature of certification. Moreover, official registers tell us little about the history and design of such buildings and their meaning to the communities that they serve. Indeed, they mask a range of 'biographies' of places of worship, some of which may be traced across the course of the twentieth century. Notwithstanding the conceptual issues that arise in the interpretation of such data, this brief review of the evolution of Muslim, Sikh and Hindu communities demonstrates that the conventional historical geographical narrative of twentieth-century Britain characterized by increasing secularization is misleading. Declining attendance at Christian churches is clearly a reality. However, this is neither an accurate measure of the extent of Christian belief nor of the growth of religious affiliation in Britain more generally. We would take this argument further, to suggest that failure to consider non-Christian religious groupings and spaces within the historical geography of twentieth-century Britain carries a political consequence. At a time when religious and ethnic divisions seem so quick to explode into life (cf. Raban 2002), it is imperative that we acknowledge and investigate the longstanding and integral place of ethnic minority and non-Christian religious groups in the landscape of modern Britain.

ACKNOWLEDGEMENTS

This chapter stems from research work on the Leverhulme-funded research project on 'Ethnicity and Cultural Landscapes' at the School of Geography, University of Oxford, 1997–2000. The project surveyed officially registered Muslim, Sikh and Hindu places of worship in England and Wales. The authors wish to extend their thanks to the project director Ceri Peach as well as to Richard Gale and other participants in the 'New Landscapes of Religion' conference held at the University of Oxford in September 2000.

REFERENCES

Amin, A., and N. Thrift 2002: Guest editorial: cities and ethnicities. *Ethnicities*, 2(3), 291–300.

Badham, P. (ed.) 1989: *Religion, State and Society in Modern Britain*. Lampeter: Edwin Mellen Press.

Beckerlegge, G. 1997: Followers of 'Mohammed, Kalee and Dada Nanuk': the presence of Islam and South Asian religions in Victorian Britain. In J. Wolffe (ed.), *Religion in Victorian Britain*. Manchester: Manchester University Press and the Open University, 221–70.

Brown, C. 1996: Review essay: religion in the city. *Urban History*, 23, 373–9.

Burton, A. 1998: *At the Heart of the Empire: Indians and the Colonial Encounter in Late-Victorian Britain*. Berkeley: University of California Press.

Crinson, M. 1996: *Empire Building: Orientalism and Victorian Architecture*. London: Routledge.

Driver, F., and D. Gilbert 1997: Heart of empire? Landscape, space and performance in imperial London. *Environment and Planning D: Society and Space*, 16, 11–28.

Dwyer, C. 1996: Contested spaces: mosque building and the cultural politics of multiculturalism. Paper presented at the Annual Conference of the Association of American Geographers, Boston, MA, Apr.

Dwyer, C. 1999: Contradictions of community: questions of identity for young British Muslim women. *Environment and Planning A*, 31, 53–68

Eade, J. 1996: Nationalism, community, and the Islamization of space in London. In B. Metcalf (ed.), *Making Muslim Space in North America and Europe*. Berkeley: University of California Press, 217–33.

Eade, J. (ed.) 1997: *The Making of Post-Christian Britain: A History of the Secularization of Modern Society*. London: Longman.

Gale, R., and S. Naylor 2002: Religion, planning and the city: the spatial politics of ethnic minority expression in British cities and towns. *Ethnicities*, 2(3), 387–409.

Gilbert, A. D. 1980: *The Making of Post-Christian Britain*. London: Longman.

Graham, B., and C. Nash (eds) 2000: *Modern Historical Geographies*. Harlow: Pearson.

Hastings, A. 1991: *Church and State: The English Experience*. Exeter: University of Exeter Press

Holloway, J. 2000: Institutional geographies of the New Age movement. *Geoforum*, 31, 553–65.

Holloway, J., and O. Valins 2002: Editorial: placing religion and spirituality in geography. *Social and Cultural Geography*, 3(1), 5–10.

Knott, K. 1986: *Religion, Identity, and the Study of Ethnic-Minority Religions in Britain*. Community Religions Research Papers. Leeds: University of Leeds.

Kong, L. 1990: Geography and religion: trends and prospects. *Progress in Human Geography*, 14, 353–71.

Kong, L. 1993: Ideological hegemony and the political symbolism of religious buildings in Singapore. *Environment and Planning D: Society and Space*, 11, 23–45.

Kong, L. 2001a: Mapping 'new' geographies of religion: politics and poetics in modernity. *Progress in Human Geography*, 25, 211–33.

Kong, L. 2001b: Religion and technology: refiguring place, space, identity and community. *Area*, 33, 404–13.

Kushner, T. 1994: Immigration and 'race relations' in postwar British society. In P. Johnson (ed.), *Twentieth-Century Britain: Economic, Social and Cultural Change*. London: Longman, 411–26.

Ley, D. 1995: 'Between Europe and Asia: The case of the missing sequoias. *Ecumene*, 2, 184–210.

Mackinder, H. 1902: *Britain and the British Seas*. London: Heinemann.

Mason, D. 2000: *Race and Ethnicity in Modern Britain*, 2nd edn. Oxford: Oxford University Press.

Matar, N. 1998: *Islam in Britain, 1558–1685*. Cambridge: Cambridge University Press.

Metcalf, B. (ed.) 1996: *Making Muslim Space in North America and Europe*. Berkeley: University of California Press.

Modood, T., R. Berthoud, J. Lakey, J. Nazroo, P. Smith, S. Virdee and S. Beishon 1997: *Ethnic Minorities in Britain: Diversity and Disadvantage*. London: Policy Studies Institute.

Naylor, S., and J. R. Ryan 2002: The mosque in the suburbs: negotiating religion and ethnicity in South London. *Social and Cultural Geography*, 3(1), 39–60.

Nye, D. 1993: Temple congregations and communities: Hindu constructions in Edinburgh. *New Community*, 19, 201–15.

Park, C. 1994: *Sacred Worlds: An Introduction to Geography and Religion*. London: Routledge.

Parsons, G. (ed.) 1993: *The Growth of Religious Diversity: Britain from 1945*, vol. 1: *Traditions*. London: Routledge and Open University

Raban, J. 2002: Truly, madly, deeply devout. *The Guardian*, Saturday Review section, 2 Mar., 1–3.

Raivo, P. 1997a: Comparative religion and geography: some remarks on the geography of religion and religious geography. *Temenos*, 33, 137–49.

Raivo, P. 1997b: The limits of tolerance: the Orthodox milieu as an element in the Finnish cultural landscape, 1917–1939. *Journal of Historical Geography*, 23, 327–39.

Rogoff, J. 1998: Sacred places: the Neasden mandir, north-west London. BA dissertation, School of Geography, University of Oxford.

Thomas, T. (ed.) 1988: *The British: Their Religious Beliefs and Practices*. London: Routledge.

Vertovec, S., and A. Rogers (eds) 1998: *Muslim European Youth: Reproducing Ethnicity, Religion, Culture*. Aldershot: Ashgate.

Weller, P. 1992: Inter-faith roots and shoots: an outlook for the 1990s. *World Faiths Encounter*, 1, 48–57

Weller, P. (ed.) 1997: *Religion in the UK: A Multi-Faith Directory.* Derby: University of Derby and the Inter-Faith Network for the United Kingdom.

Wolffe, J. (ed.) 1993: *The Growth of Religious Diversity: Britain from 1945, A Reader.* London: Hodder & Stoughton

Wolffe, J. 1994: Religion and 'secularization'. In P. Johnson (ed.), *Twentieth-Century Britain: Economic, Social and Cultural Change*, London: Longman, 425–41

Yeoh, B., and T. B. Hui 1995: The politics of space: changing discourses on Chinese burial grounds in post-war Singapore. *Journal of Historical Geography*, 21, 184–201.

Part III

Geography, Nation, Identity

Chapter Ten

'Stop being so English': Suburban Modernity and National Identity in the Twentieth Century

David Gilbert and Rebecca Preston

I speed up again...past what were once known as stockbrokers' Tudor houses, although there are no stockbrokers here and there never were. Instead there are software buyers, computer programmers, local shopkeepers, skilled workers, financial services salesmen, bank employees. These houses...look worn and dispirited. It is hard to believe that they were built as part of a great common dream of England...They *are* England, more than any palace or flag, with their dull, decent reflection in every city and town. (Lott 1996: 26)

Who wants to have an identity stamped Made in the Suburbs? But when I finished...my adventures in suburbia, I went out and bought some nice white shoes. And wore them with some pride. Because Britain was being built by its suburbs and my past was shaping the present. And because I thought they might make me look taller. (Sawyer 1999: 309)

Suburbia has good claim to be the locus of twentieth-century modernity, particularly in North America, Australia and Britain. In his introduction to a landmark collection of essays, Roger Silverstone draws attention to this. Most accounts of modernity have given central place to the urban, most obviously and brilliantly that by Marshall Berman (1982), but for millions of people 'the experience of modernity was the experience not of the street, but of the road, not the sidewalk but the lawn, and not the jarring and unpredictable visibility of public spaces and public transport, but the enclosed private worlds of fences, parlours and automobiles' (Silverstone 1997: 5). Silverstone's collection, *Visions of Suburbia,* has been part of a wider intellectual movement that has reinterpreted and revalued suburbia (see e.g. Oliver et al. 1981; Gold and Gold 1989; Carey 1992; Webster 2000). To some extent this work sits in a longer tradition challenging the established stereotypes of suburbia and suburban life. The emergence of modern suburbia in the nineteenth century was accompanied very rapidly not just by a standard set of disparaging tropes (which have remained remarkably persistent to the present day), but also by counter-accounts that emphasized both the positive qualities of the suburbs and their more ambiguous hidden depths. But what might be described as 'the new suburban studies' go further than this. Instead of seeing suburbia as a monolithic and monotonous counterpoint to the diversity and energy of the modern city – as literally less than urban – these studies show that the conceptual apparatus used to understand the supposedly distinctively urban dimensions of modernity can also be applied to suburban spaces (Ryan 1995). Suburbia may not be as focused or as obviously spectacular as the city, but ideas of creative destruction, ambiguity, doubled-sidedness and fragmented experience have very direct relevance to its interpretation.

One dimension of this approach relates very directly to the geographies of suburbia. A common, perhaps dominant, account of suburbia has emphasized its placelessness. Ironically, this view of suburbia as 'nowhere' has been applied to strikingly different places. In Britain this has included Victorian 'villadom', late nineteenth-century and Edwardian terraced suburbs, inter-war semi-detached suburbs, New Towns and suburban public estates of the 1950s, and the owner-occupied estates of the building booms of the 1980s and 1990s. At different times particular places came to represent this placelessness: the spec-build suburb of Camden Town (as satirized in cartoon form by George Cruikshank as early as 1829), the Holloway of 'The Laurels' (in the Grossmiths' *Diary of a Nobody* of 1892), and inter-war Slough (as in Betjeman's 'Come friendly bombs'). At the beginning of the twenty-first century, Swindon (or more precisely the new-build 'executive' estates around Swindon) seems to have become the new English epitome of nowhere. One characteristic of the term 'suburban' is the slipperiness of its relationship with its signified. The word is so over-burdened with cultural meanings that it is rarely used as a simple description of urban morphology. In twentieth-century England, almost anywhere outside the Circle Line could become nowhere – in the planner Thomas Sharp's phrase (1936), part of a 'universal suburbia'.

This trope of placelessness was particularly well developed in the work of critics of 1950s and 1960s American suburbia, reaching maturity in Lewis Mumford's work. In *The City in History* (1961) the suburb is treated as a betrayal of thousands of years of city-building, as an act of erasure of urban distinctiveness. There are continuities here with the currently widespread adoption of Marc Augé's (1995) concept of 'non-place' as an element of the experience of 'supermodernity'. Alongside airports, motorways, super-markets and shopping malls, the landscape of suburbia is presented as detached from both local particularity and historical time. Theorists of post-modern urbanism, notably in the 'Los Angeles school', may emphasize the differences between a late twentieth-century landscape of theme parks, giant malls and edge-city, and the 'Fordist' suburbia of mid-twentieth-century America (for an overview see Dear and Flusty 1998), yet the underlying conception of the suburban is consistent: the homogenizing forces of capitalism have created non-places stripped of meaning and reference.

The new suburban studies can be seen as a rejection of this portrayal of the suburbs as lacunae. They achieve this in various ways. One method is through an emphasis on the cultural forms of suburbia, emphasizing individual creativity and the complexity of consumption practices. Studies of, for example, the role of suburbia in the development of English popular music (Frith 1997; LeBeau 1997) or the influence of Tupperware parties on identity and associational culture in 1950s America (Clarke 1997) can be seen as a part of a broader effort to establish the particularities of suburban history. This is not to deny that suburbia (like just about every other landscape on the planet) has been subject to the influence of global capitalism, and that this influence has often had a strongly homogenizing effect. It is rather to suggest that suburbia, every bit as much as the city, is characterized by people making their own history (sometimes in this case in circumstances rather precisely of their own choosing). A second response is to think about suburban sites as specific points in wider flows and connections. Again this way of thinking has approached orthodoxy in urban studies and human geography more generally. 'World cities' are understood and explained primarily in terms not of their internal characteristics, but rather their place in global networks of monetary and commodity exchange, political power, human migration, and symbolism and meaning. Using the example of the ordinary Inner London district of Kilburn, Doreen Massey has shown the way in which its nature as a place is constructed and contested through its position in networks that stretch far beyond its boundaries (Massey 1991, 1995). But this kind of perspective has too rarely been applied to suburbia. There have been some attempts to think about the place of English suburban homes in the networks of empire, unravelling what Raphael Samuel described provocatively as the 'English diaspora' (Samuel 1998: 74ff; for other examples of work that addresses English suburbia as an imperial space see Ryan 1999; Preston 1999; Blunt 1999). Anthony King has used the 'hidden histories of the bungalow' as an alternative route through the transnational geographies of the suburban. This goes far beyond the simple recognition of the stylistic and etymological origins of the bungalow in India to considering the marketing of 'pure American-Canadian' bungalows in the new suburbs of late twentieth-century Beijing (King 1997: 81).

A further approach to the placing of suburbia recognizes the extent to which it became intertwined with national identity over the course of the twentieth century. This supposedly anodyne and anonymous landscape has often become the key site for debate about the nature of the 'imagined community' of many Western nations. John Updike's *Rabbit* tetralogy, perhaps the most perceptive fictional portrayal of post-war American identity, is set, not in New York or Los Angeles, but in Brewer, Pennsylvania. The key moments and movements of American life – consumer affluence, race relations, women's liberation, pop culture, Vietnam, Nixon, the moon landings, Reaganomics – are seen through the eyes of an increasingly sclerotic and overweight suburban car salesman. Somewhere between Frank Capra's *It's a Wonderful Life* (1946) and Steven Spielberg's *ET* (1982), Hollywood's preferred topographical signifier for ordinary American life also shifted from the small town to the suburb.

In Britain too, suburbia has often been the reference point both for cultural constructions of ordinariness and in debates about the state of the nation(s). Despite the existence of substantial areas of suburban development around Cardiff, Edinburgh, Glasgow and Belfast, the suburban has been primarily associated with Englishness. This division has a reinforcing momentum. Often constructions of Welshness, Scottishness and Irishness have been defined in opposition to the idea of an effete suburban Englishness. The cultural histories of Welsh, Scottish, and varieties of Irish suburbia remain relatively hidden and require further research, but this chapter concentrates its attention on the uneasy relationship between suburbia and Englishness. It is not intended to be a comprehensive review; instead it will approach the issue from three directions. The following section provides a brief synopsis of what might be described as English anti-suburban rhetoric. As has already been suggested, while the physical form, social structure and geographical location of what was described as suburban changed significantly through the twentieth century, the disparagement of suburbia remained remarkably consistent. Running through this rhetoric has been a double-sided concern for national identity, an ambiguity that sometimes presents suburbia as the quintessence of Englishness, and at other times has portrayed it as a betrayal of England. The chapter then looks at this connection between suburb and nation by examining the political representations of late twentieth-century national leaders, particularly Margaret Thatcher and John Major. The final section of the chapter examines the late twentieth century, when the idea of a crisis in suburbia itself was used as a symbol of national decline.

Suburbia and Englishness

In his recent book on English identity, Robert Colls has claimed that 'most anti-suburban prejudice was pre-1914', and that by the 1930s the 'well-swept avenues and striped green lawns came to be seen as the heart of England itself' (Colls 2002: 217–18). While it was undoubtedly the case that the great expansion in suburban England which took place between the wars helped to add the 'invincible green suburb' to a list of archetypal national landscapes, if anything this growth created an intensification of anti-suburban prejudices rather than a relaxation. These were most marked, or perhaps just most vividly expressed, in parts of the 'literary intelligentsia' and in the architectural and planning professions (see Carey 1992; Oliver et al. 1981; Gold and Gold 1989; Matless 1999). Commenting on the attitudes of the intellectual elite in the 1930s, John Carey argued that the term suburban was 'distinctive in combining topographical with intellectual disdain', relating 'human worth to habitat' (Carey 1992: 53). This correspondence between people and place continued in the post-war period, most notably in Ian Nairn's 1955 definition of an idealized suburbia as 'Subtopia': 'Visually speaking, the universalization and idealization of our town fringes. Philosophically, the idealization of the Little Man who lives there' (Nairn 1956: 365). The same rhetoric survives into the early twenty-first century, notably in the work of architectural writers like Jonathan Glancey and Jonathan Meades. Meades recently revived the term 'subtopia', 'an ugly coinage for an ugly and ubiquitous sort of place whose multiplication is apparently unstoppable' (Meades 2002).

Contested ideas of Englishness were very rarely far from the surface in this kind of writing. Throughout the twentieth century, suburbia was castigated both for being too English and for not being English enough. The 'too English' line often emphasized a loss of national vitality or England's inability to be sufficiently modern. Suburbia and suburbanites were seen as symptomatic of a nation that was losing its sense of destiny, and was turning in on itself. For writers like Eliot, Auden or Lawrence, England had become too domestic, too safe, and inspiration needed to be sought abroad, away from the suffocations of 'home' (see Fussell 1980). Alison Light (1992) has suggested this rejection of 'home' elided nation and the domestic landscape of the new inter-war England, and was strongly gendered. For many male writers who saw themselves as a part of an international literary culture, suburbanized England was an increasingly emasculated space, a nation dominated by the humdrum petty demands of everyday life, populated by henpecked 'little men' and their dominant wives. The response to these changes was often vitriolic misogyny, as in D. H. Lawrence's description of suburban houses as 'horrid little red mantraps' (quoted in Oliver et al. 1981: 11), and expressed most graphically in George Orwell's *Coming Up For Air*, in which the typical inhabitant of a 'line of semi-detached torture chambers' is beset by 'the boss twisting his tail and his wife riding him like the nightmare and his kids sucking his blood like leeches' (Orwell 1959 [1939]: 14).

Architectural critics of suburbia shared this disdain for the 'too English'. In the 1930s the speculatively built semi-detached house became emblematic of all that was wrong with English architecture, explicitly contrasted with 'international' modernism. To the architectural establishment the semi reeked of timidity, compromise and profit-centred pandering to the uneducated tastes of the lower middle classes. Stylistically, it was the worst kind of historical kitsch, with stuck-on references to earlier periods of vernacular design. This reached its apotheosis in what the architect and preservationist John Gloag described as the 'half-baked pageant' of the 'tudorbethan' style with its fake half-timbering (Gloag 1934: 20). While these symbolic echoes of England's Tudor glories may have provided reassuring comforts for their owners, they were the strongest of provocations to the architectural and planning professions. By the late twentieth century the attack on the new-build house was often couched in the specific language of 'design' but the themes remained the same. Accusations of a feminization of culture that echo Lawrence and Orwell are to be found in A. M. Edwards's 1981 book *The Design of Suburbia*, where he speculates on causes of the 'ugliness' of suburbia, 'hideous in appearance and so alarming in its spread', and concludes that 'the influence of female emancipation on architectural taste has never been studied but it must have been considerable' (Edwards 1981: 130, 134). The out-of-Swindon estate is now held up as the exemplar of national bad taste, and of a collective failure of imagination and creativity. Jonathan Glancey's polemic in 2002 against this 'first horror of New Britain' uses terminology and imagery almost unchanged from inter-war rhetoric: 'the executive housing that grasps its tentacles around each and every town, smothering them with kitsch design, improbable mortgages, company cars and cul-de-sacs'.

An alternative view of suburbia held that it was not English enough, a desecration of a true England – although many critiques contained contradictory elements of both views. David Matless (1999) has argued that the mid-twentieth century was characterized by an alliance between planners and preservationists which sought to reconcile progressive visions of a new England with the conservation of older landscape elements. The key to this was a very modern concern for function and order that sought strong boundaries between the urban and the rural. From this perspective, suburbia could be a betrayal of both England's past and its future. From the 1920s onwards, particularly with the growing impact of the car on English suburbia, one take on this argument saw it as a beachhead for an American invasion. In his *English Journey*, J. B. Priestley identified an emergent Americanized 'Third England' to set alongside the countryside and the industrial towns: 'essentially democratic in its cheapness', this was an England of 'arterial and by-pass roads', 'giant cinemas' and 'bungalows with tiny garages' (Priestley 1934: 401). Priestley was not unambiguously hostile to this new world of the lower-middle classes, but laments about the Americanization of England became ever more common during the second half of the twentieth century. The criticism applied not only to the physical landscape of suburban houses, shopping malls, cinema multi-plexes and commercial strip developments along main roads, but also to lifestyles, where a supposedly English culture of politeness, community-spiritedness and quietude (ironically often seen as suburban characteristics) was being usurped by a rougher, more selfish and privatized American consumerist culture.

Critical readings of the connections between national and suburban identities did not go unchallenged. Another strain of writing saw the suburbs as the natural habitat for certain kinds of English virtue and welcomed the rise of the world of the 'little people'. While Orwell could provide the most vituperative castigation of the spec-build suburb, he also wrote about Englishness in ways that clearly included the suburban world as well as older industrial and rural communities: 'another English charac-teristic which is so much a part of us that we barely notice it, and that is the addiction to hobbies and spare-time occupations, the privateness of English life. We are a nation of flower-lovers, but also a nation of stamp-collectors, pigeon-fanciers, amateur carpenters, coupon-snippers, darts-players, crossword-puzzle fans' (Orwell 1962 [1941]: 13). Similar contradictions appeared in the work of John Betjeman, remembered equally for his call for 'friendly bombs' to fall on Slough in his poem of 1937 and the extreme suburbophilia of his 1973 BBC television eulogy to Metroland.

In general, however, the argument for suburbia as a positive development in national identity was not associated with major literary figures, but with the less well known, often suburbanites themselves – *People Like Us* – as in the collective title of R. F. Delderfield's suburban 'saga' set between 1919 and 1947 in a fictional south London suburb. First published in 1958, the saga reached a wider audience in 1978 with a television serialization. *The Dreaming Suburb* and *The Avenue Goes to War* are 'an account of the lives and dreams of five families of the Avenue' (Delderfield 1978: 2), a tree-lined street of terraced houses built shortly before the First World War. For Delderfield, these suburban streets are the true 'sources' of history, their inhabitants 'the British thought-developers and policy makers' who collect-ively exhibit the 'trends and thought and emotion that ultimately become the policy of a nation' (1978: ix). Social development – 'what most of us recognize as progress' – originates chiefly in the suburbs. He laments the fact that suburban people are 'for the most part unsung…even though they represent the greater part of Britain's population. The story of the country-dwellers, and the city sophisticates, has been told often enough; it is time somebody spoke of the suburbs, for therein, I have sometimes felt,

lies the history of our race' (1978: v). As Colls suggests, the new suburbanites of the 1930s 'liked their airy modern houses and they made money on
them. They certainly were not going to accept that there was anything
un-English about where they lived' (Colls 2002: 218). Instead, what was
being created was a new inflection of Englishness which was capable of
accommodating dramatic changes in social and physical landscapes within
a rhetoric of continuity – what Alison Light (1992) has characterized as
'conservative modernity'.

However, this is not to imply that there was a general pro-suburban
consensus among English suburbanites. Rather, middle-class identities
often depended on 'an extremely anxious production of endless discriminations between people who are constantly assessing each other's standing'
(Light 1992: 13). In these endless discriminations, the combination of
social and topographical disdain was extremely significant. Lynne Habgood
(2000) has suggested that the emergence of anti-suburban tropes in popular journalism (particularly in the Grossmiths' columns for *Punch*) had less
to do with intellectuals' fears about mass society than with intense cultural
conflict between the established and the newly emerging middle classes.
The humour in *Diary of a Nobody* works by allowing its audience to
simultaneously empathize with and keep its distance from Mr Pooter's
manners and circumstances. Since Pooter there has been a long and
developing tradition of popular culture in England that has used this
tension to great effect. As Andy Medhurst (1997) has argued in an essay
on 'negotiating the gnome zone', this has almost always worked through
comedy, in plays, novels, film and television. In this context the negoti-
ations are between positive representations of suburbia as decent and down
to earth and negative representations of suburbia which recast many of the
elite critiques into particular characters and situations.

Two late twentieth-century advertising campaigns for the Swedish furniture retailer Ikea, purveyor of flat-pack modernism to the masses, indicate
this lasting tension. A TV advertisement from 1997 called upon home-
owners to 'chuck out their chintz'. In an archetypal bay-fronted suburban
street, a crowd urges people to defenestrate their frilly patterned sofas,
armchairs and lacy curtains, in an act of soft-furnishing reformation. The
follow-up campaign reinforced the point: consumers were urged to 'Stop
being so English'. Not only were chintz, lace and frills indelibly associated
with the 'Terry and June' world of humour-free TV situation comedy, but
they were also indicative of English backwardness and provincialism in the
face of Scandinavian modernity. As Stephen Moss later noted, it was a
happy coincidence that these adverts appeared in 1997, in the euphoria of
the Blairite new dawn. Ikea called upon the English to modernize, to
become 'tough on chintz and tough on the causes of chintz... like New
Labour, offering us a new beginning: clarity, classlessness, opportunity, a

bright future' (Moss 2000). But for all its surface appearance of a revolutionary change in English interior design, Ikea's advertising was working within well-established understandings of suburban culture and the place of the modern within it. This advertising was not calling people to leave suburbia and to become metropolitan sophisticates – rather, it was encouraging the use of modern design as a marker of distinctions within suburban consumer culture. The knowing irony of the campaigns was very 1990s, but the basic premise was the same as that used to sell cheap Deco furniture, or Ercol and G-Plan in the past.

In Search of Middle England: Thatcher, Major and Blair as Suburban Leaders

The advent of universal suffrage between 1918 and 1928 helped to create what some have described as a 'domestication' of party politics during the 1930s (see Pugh 2000; Beddoe 1989). The extension of the vote did not create the radicalization of British politics that many had expected or feared. Rather than providing a boost to a specifically feminist politics, women's suffrage had the effect of pushing forward what might be described as a suburban agenda. Under Stanley Baldwin, the Conservative party retreated from its traditional role as the defender of crown and empire towards the politics of housing, cost of living, education and health. Unlike in the late nineteenth century, when the politician had 'no place for the suburb in his programme' and could be 'half ashamed' of electoral victory in a suburban constituency, the suburban was moving to centre stage in British politics (Low 1891: 558).

By the late twentieth century British political life had developed a keen popular sense of the geographical location of swing voters and marginal constituencies. During the Thatcher years, Basildon, with its population of 'Essex men' and 'Essex girls' held a talismanic significance for the Tory party. Through council house sales and an appeal to what Stuart Hall described as a combination of 'reactionary common sense and authoritarian populism' (Hall 1983: 30), the Conservative party seemed for a while to have turned the Basildons of Britain into their natural territory. In the 1990s, with the advent of New Labour, the electoral battleground of British politics shifted to the mythical kingdom of 'Middle England' – a land that could never quite be identified precisely on the map, but which seemed to include most places that were outside the cities but not in the countryside (see chapter 3 above). In an electoral system where swings between parties are crucial, places which tend to switch allegiance come to be regarded as indicative of broad changes in national characteristics. It is the suburban landscape, and the wider English edge-city, that has come to hold this role.

Another striking feature of late twentieth-century politics was the way in which television focused attention on the personality and lifestyle of the prime minister. This, of course, was part of a longer process. Baldwin, for example, had used radio and the popular press to cultivate a public persona of honesty and ordinariness that drew upon his geographical roots in the deep England of Worcestershire. But this kind of personalization of politics was greatly heightened by the immediacy of television. From Harold Wilson onwards, the domestic life of the premier became part of the cultural currency of mass politics, and party image-makers were keen to present leaders as being 'of the people'. Suburban landscapes and culture became important markers of ordinariness, and of the leader's embodiment of national identity.

Margaret Thatcher's relationship with suburban Englishness was more complicated than might at first appear. In the early years of her leadership of the Conservative party, she was frequently presented as, if not quite a domestic goddess, then certainly some kind of *über*-housewife. The Thatcher home (at that time a decidedly upmarket London house just off the King's Road in Chelsea) became a metaphor for the nation. The complexities of supply-side economics were presented as domestic science, through 'the homilies of housekeeping or the parables of the parlour' (speech at the Lord Mayor's Banquet, November 1982, quoted in Riddell 1985: 7). Part of the reaction to Britain's first female prime minister worked through this representation, using a potent mixture of class snobbery, misogyny and anti-suburban rhetoric. Ian Gilmour, from the patrician end of the Conservative party, expressed his contempt on Thatcher's accession to party leadership in these terms: 'We cannot really believe that this is the moment for the party of Baldwin and Churchill, of Macmillan and Butler, of the Industrial Charter and the social advances of the 1950s to retreat behind the privet hedge into a world of narrow class interests and selfish concerns' (quoted in Gamble 1988: 140). From a very different political perspective, Dr Jonathan Miller used similar language, criticizing Thatcher's 'odious suburban gentility and sentimental, saccharine patriotism, catering for the worst elements of consumer idiocy' (quoted in Evans 1997: 118).

In many ways the rise of Thatcherism did feel like a lower-middle-class revolution, with privet hedges for barricades. Conservative party confer-ences in the 1980s seemed to be filled with Thatcher wannabes – a braying sea of hats and blue-rinsed hair, fitting closely to stereotypes of suburban 'little Englanders'. Government policy was designed to create a nation of privatized home-owners, extending the boundaries of Conservative-voting England into unknown territory. The sale of council houses was a potent form of social gerrymandering. Home ownership increased from 54 per cent of dwellings in the UK in 1979 to 67 per cent in 1990, with around half of the increase as the direct result of 'right-to-buy' schemes (Williams 1992: 166). The liberalization of the mortgage market and the extension of share-ownership associated with the sale of public assets further heightened the sense of a suburban nation on the march.

Yet the Thatcherite Conservative party misjudged the character of what it took for its homeland. The party was marked by a disdain for the social (there was after all 'no such thing as society' as Thatcher stated in *Woman's Own* in 1987), and by a fanatical belief in brute economic logic that was as strong as anything the head-banging end of historical materialism could come up with. Thatcher herself, although inclined to stress the virtues of the voluntary sector in the abstract, was famously disinterested in the everyday organization of her own party. For someone who increasingly saw herself as a world leader with a strong sense of her manifest destiny, the parochial business of social clubs, garden fêtes, raffle tickets and tombolas seemed a petty irrelevance. But until the 1980s the Conservative party had been more than the passive recipient of suburban votes; it had gained strength through its role as one of the cornerstones of suburban associational culture. In the 1980s the party shifted from a mass social movement characterized by a certain disinterest in politics towards an ideological sect, and its membership, which had been over 2.5 million in the early 1960s, declined ever more rapidly. By the early 1990s, membership had fallen to a half a million, and half of the members were over 65 (Seyd and Whiteley 1996: 72).

In 1987 the Thatchers bought a new house in suburban south London, at 11 Hambledon Place, just off the South Circular Road in Dulwich, close to Sydenham golf club. They paid over £400,000 for a top-of-the-range Barratt home. It ought to have helped connect Mrs Thatcher to the population. But the Dulwich residence became a symbol not of ordinariness but of aloofness, detachment and national division. The Barratt development was an early British example of a gated community, fully enclosed by high walls, and equipped with closed-circuit television and electronic entrance gates. This may have made good practical sense for the most widely hated prime minister of modern times, but the symbolism was disastrous. Not only did the Dulwich gates emphasize Thatcher's disconnection from everyday life, but they also symbolized the intrusion of American-style social polarization into English suburbia. It was an impression that was reinforced when she turned Downing Street into a gated community in 1989.

Margaret Thatcher's successor has some claim to be the first truly suburban prime minister. Others had come from the middle classes, but John Major was the first from the deep heartlands of the gnome zone. Major's childhood family home was in Worcester Park, classic Outer London spec-build suburbia. In his autobiography, Major remembered the family home:

a small bungalow with four rooms, a bathroom and a kitchen. Our garden was long and narrow, dotted with sheds in which my father worked. We had a lawn just large enough for ball games, and two ponds: one shallow with a few goldfish, the other a deep iron tank sunk into the ground. There were rockeries, fruit trees to plunder and larger trees to climb. (Major 1999: 8)

John Major's early years were suffused with suburban culture. It is well known that his father had worked as a travelling trapeze artist, but he had settled down in Worcester Park long before John was born. In the early 1930s Tom Major had set up his own business, making garden ornaments, gnomes included. Major's Garden Ornaments prospered with the expansion of semi-detached London diversifying into crazy paving, turfing and landscape gardening. The business went bust in 1955 (more to do with some shaky financial arrangements than lack of demand for the accoutrements of suburban gardening), and the family moved to a small rented flat in Brixton. John Major spent his adolescence in Brixton, and it was there that he started his career in Conservative politics.

This childhood displacement from deep suburbia to the inner city allowed different stories to be told of Major's rise. In the early years of the Major premiership, the story of the Brixton boy was promoted strongly. It trumped some of the common opposition criticisms of Tory leaders: here was a leader not from the aristocracy or the shires, but from the inner city. But there was an alternative narrative, one that emphasized Major's ordinariness and his suburban origins. The former Tory Whip, Tristan Garel-Jones, described Major as the 'personification of Middle England': 'When my constituents ask what he's like, I say he's the sort of person I would expect to see with his car parked by the pavement on a Sunday, washing the car, eating some Polo mints, and listening to the cricket match on the radio. He is extraordinarily ordinary' (quoted in Junor 1996: 112).

It was a portrait that Major himself seemed to be comfortable with. Much of the rhetoric of Major's social policy looked back to an idealized 1950s Worcester Park, particularly in the ill-fated 'back-to-basics' campaign (fatally undermined by the predilection of Tory MPs for backhanders and sexual shenanigans, and rendered more risible by later revelations about Major's own personal life). In one of his best-remembered (and most ridiculed) speeches, he expressed his love for an England of 'long shadows on county grounds, warm beer, invincible green suburbs, dog lovers and old maids bicycling to Holy Communion through the morning mist'. Although the words were drawn from Orwell, the sentiment was Baldwin's. Baldwin too had used the rhetoric of deep England to appeal to the conservative aspirations of suburbanites. Major tried the same trick – except that the site of prime ministerial experience had shifted from Baldwin's Worcestershire to the somewhat green and decidedly vincible suburb of Worcester Park.

There were, however, distinct limits to 'likeability' (Broughton 1999: 200). If John Major's suburban ordinariness had been seen as a significant strength in the early years of his premiership, after the debacle of Britain's ejection from the exchange-rate mechanism it was increasingly seized upon as evidence of his weakness and lack of leadership qualities. As the standing of his administration plummeted, Major's lower-middle-class suburban roots and attitudes became the focus of satire and mockery. There were many variants on Prime Minister Pooter. In *Private Eye* 'The Secret Diary of John Major Aged 47¾'crossed Adrian Mole with an updated version of *Diary of a Nobody*. On television's *Spitting Image*, a totally grey Major puppet droned on at his wife while desultorily picking at a plate of luke-warm peas. Major's wardrobe of plain grey suits and Marks & Spencer's pullovers didn't help matters, nor did occasional stops for meals in branches of Happy Eater. His background and manner may not have been the cause of his government's unpopularity, but as his position weakened the full force of a long-established tradition of anti-suburban rhetoric was turned against him. Tristan Garel-Jones's supposedly affable description of the Polo-sucking premier betrayed underlying elitist preju-dices about the suitability of suburbanites in power.

In contrast with his predecessor, Tony Blair lacked a strong public connection to place. There were associations with the metropolitan sophis-tication of Islington, his north-eastern constituency of Sedgefield, and earlier student connections with Edinburgh. However, when compared with John Major's competing myths of origin and pronounced London suburban accent, Blair seems a kind of human 'non-place', a nowhere man, lacking clearly identifying geographical characteristics. He was of course, the man that made it safe for suburban England to vote for the Labour party, and it is telling that, in the search for electability, New Labour's strategists shifted from people in places to people in cars. In place of 'Essex man' came 'Mondeo man', 'Sierra man', and 'Galaxy man', target voters who were almost by definition creatures of the suburban world, but not of any specific point within it. Blair himself is most identified not with a gated executive home, 1920s semi, or Brixton flat, but with his car. It seems no coincidence that his vehicle of choice is a people carrier, and not a humble Sharan, but a Grand Voyager at that.

Coda: Suburbia and Narratives of English Decline

The closing years of the twentieth century were marked by a new inflection in the relationship between suburbia and Englishness, one that interpreted suburban decline as a bellwether of national decline. A Rowntree Trust report of 1999 (Gwilliam et al.) suggesting that some suburban areas were showing signs of stress was greeted with surprise. Here was a story not of invincible green suburbs, but of suburbs in decay. This theme was not completely new. Critics of suburban development in the mid-twentieth century had suggested that they would become the 'slums of the future' (see Lancaster 1938), while eulogies such as James Kenward's *The Suburban Child* lamented that the 'great days' of suburbia 'are already over' (1955: 1). But for the first time the Rowntree report consolidated empirical research about failing infrastructure and declining housing stock in suburban areas. Given the connections established between suburbia and Englishness over the previous century, it was unsurprising that this was taken as evidence of national decline.

The connection between Englishness and suburban decline found its most profound expression in Tim Lott's autobiography *The Scent of Dried Roses* (1996). Lott tells the story of his mother Jean's suicide and his own depression through a wider narrative of declining suburbia. Born and brought up in Southall, west London, Lott tells a story of lost hopes and an increasing sense of anomie among the generations that had first moved into the new suburb:

> Somewhere in this hall…I feel a resonance of the England Jean perhaps imagined she wanted to belong to – warm, slightly eccentric, innocent, quiet, decent, quaint, a bit pompous, fond of a lark. It is probably the same obscure, half-inherited feeling I get watching *A Matter of Life and Death*, *Colonel Blimp*, *The Ladykillers* or *Arsenic and Old Lace*. It is the sense of an impossible, gentle, romantic imagined England, an alternative to the pompous, conceited, bullying Imperial England hymned still by the politicians and nostalgists, or the irredeemably guilty, racist, rapacious England imagined by the marchers, the shouters, and the air-punchers. Anyway, it is an England that has gone now, even as an idea, a dream. There is nothing we have thought of to put in its place. There is, in fact, no place to put it; the cohering forces themselves have collapsed. The centre could not hold, had gone even as I was born. (Lott 1996: 34)

Much of Lott's narrative could have been set in any of a hundred English suburbs – a story of transformation from a world of pristine semis and immaculate lawns to one of concreted-over front drives, thermoplastic-framed double glazing, stone cladding, satellite dishes, raddled hedges and dog mess (Lott 1996: 4–5). In the Penguin edition this picture is reinforced by the cover illustrations, with front and back showing 'before and after' images. But Lott's combination of topographical, personal and intellectual despair was set in the very specific context of the development of Britain's largest centre of post-war South Asian settlement. Lott's history is not a simple story of defensive racism in the face of a 'swamping' invasion of immigrants. Rather it is a more complex narrative of the disruption of a world-view where whiteness and Englishness were taken-for-granted threads in the fabric of everyday life.

There is an alternative reading of this history, one that takes the idea of 'creative destruction' from interpretation of the urban to that of the suburban. The Rowntree report pointed not just to the stresses in some parts of suburbia, but also to its continuing capacity to adapt (albeit without too much fuss or financial outlay: in marked contrast to grand schemes for an 'urban renaissance', the Civic Trust rather appropriately produced a 'toolkit for suburbia'). While academic attention focuses on the social dynamism and potential hybridity of the city (often in assumed contrast with defensive, conservative suburbia), English suburbs have seen dramatic social and cultural change in the past 20 years. It is telling that a mainstream BBC television comedy show, Sanjeev Bhaskar's *The Kumars at No. 42*, is able to draw its humour, not from the incongruity of Asians in 'white' suburbia, but from the increasingly widespread knowledge of the mores of British Asian suburban life. There may be no centre, no singular dream of England, but the suburbs remain pivotal sites in social and cultural change.

REFERENCES

Augé, M. 1995: *Non-Places: Introduction to an Anthropology of Supermodernity*, trans. John Howe. London: Verso.

Beddoe, D. 1989: *Back to Home and Duty: Women between the Wars 1918–39*. London: Pandora

Berman, M. 1982: *All That Is Solid Melts To Air*. London: Verso

Blunt, A. 1999: Imperial geographies of home: British women in India, 1886–1925. *Transactions of the Institute of British Geographers*, 24, 421–40.

Broughton, D. 1999: The limitations of likeability: the Major premiership and public opinion. In P. Dorey (ed.), *The Major Premiership*. Basingstoke: Macmillan, 199–216.

Carey, J. 1992: *The Intellectuals and the Masses: Pride and Prejudice among the Literary Intelligentsia, 1880–1939*. London: Faber & Faber.

Clarke, A. J. 1997: Tupperware: suburbia, sociality and mass consumption. In R. Silverstone (ed.), *Visions of Suburbia*. London: Routledge, 132–60.

Colls, R. 2002: *Identity of England*. Oxford: Oxford University Press.

Dear, M., and S. Flusty 1998: Postmodern urbanism. *Annals of the Association of American Geographers*, 88, 50–72.

Delderfield, R. F. 1978 [1958]: *People Like Us*, vol. 1: *The Dreaming Suburb*. London: Coronet.

Edwards, A. M. 1981: *The Design of Suburbia*. London: Pembridge Press.

Evans, E. J. 1997: *Thatcher and Thatcherism*. London: Routledge.

Frith, S. 1997: The suburban sensibility in British rock and pop. In R. Silverstone (ed.), *Visions of Suburbia*. London: Routledge, 269–79.

Gamble, A. 1988: *The Free Economy and the Strong State: The Politics of Thatcherism*. London: Macmillan.

Glancey, J. 2002: Journey into an urban heart of darkness. *The Guardian*, 13 Aug.

Gloag, J. 1934: *Design in Modern Life*. London: Allen & Unwin.

Gold, J., and M. Gold 1989: *Outrage* and righteous indignation: ideology and imagery of suburbia. In F. W. Boal, *The Behavioural Environment: Essays in Reflection, Application and Re-evaluation*. London: Routledge, 163–81.

Gwilliam, M., C. Bourne, C. Swain and A. Pratt 1999: *Sustainable Renewal of Suburban Areas*. York: Rowntree/YPS.

Habgood, L. 2000: 'The new suburbanites' and contested class identities in the London suburbs 1880–1900. In R. Webster (ed.), *Expanding Suburbia: Reviewing Suburban Narratives*. Oxford: Berghahn, 31–50.

Hall, S. 1983: The great moving right show. In S. Hall and M. Jacques, *The Politics of Thatcherism*. London: Lawrence & Wishart.

Junor, P. 1996: *John Major: From Brixton to Downing Street*. Harmondsworth: Penguin.

Kenward, J. 1955: *The Suburban Child*. Cambridge: Cambridge University Press.

King, A. 1997: Excavating the multi-cultural suburb: the hidden histories of the bungalow. In R. Silverstone (ed.), *Visions of Suburbia*. London: Routledge, 55–85.

Lancaster, O. 1938: *Pillar to Post*. London: John Murray.

LeBeau, V. 1997: The worst of all possible worlds. In R. Silverstone (ed.), *Visions of Suburbia*. London: Routledge, 280–97.

Light, A. 1992: *Forever England: Femininity, Literature and Conservatism between the Wars*. London: Routledge.

Lott, T. 1996: *The Scent of Dried Roses*. London: Viking.

Low, S. 1891: The rise of the suburbs: a lesson of the Census. *Contemporary Review*, 60, 545–58.

Major, J. 1999: *John Major: The Autobiography*. London: HarperCollins.

Massey, D. 1991: A global sense of place. *Marxism Today*, June, 24–9.

Massey, D. 1995: Places and their pasts. *History Workshop*, 39, 182–92.

Matless, D. 1999: *Landscape and Englishness*. London: Reaktion.

Meades, J. 2002: Death to the picturesque! *Open Democracy*, 24 July, <www.open Democracy.net>.

Medhurst, A. 1997: Negotiating the gnome zone: versions of suburbia in British popular culture. In R. Silverstone (ed.), *Visions of Suburbia*. London: Routledge, 240–68.

Moss, S. 2000: The gospel according to Ikea. *The Guardian*, G2 section, 26 June, 1–3.

Mumford, L. 1961: *The City in History: Its Origins, its Transformations, and its Prospects*. New York: Harcourt Brace Jovanovich.

Nairn, I. 1956: *Outrage*. London: Architectural Press.

Oliver, P., I. Davis and I. Bentley 1981: *Dunroamin: The Suburban Semi and its Enemies*. London: Barrie & Jenkins.

Orwell, G. 1959 [1939]: *Coming Up For Air* London: Secker & Warburg.

Orwell, G. 1962 [1941]: *The Lion and the Unicorn: Socialism and the English Genius*. London: Secker & Warburg.

Preston, R. 1999: 'The scenery of the torrid zone': imagined travels and the culture of exotics in nineteenth century British gardens. In F. Driver and D. Gilbert (eds), *Imperial Cities: Landscape Display and Identity*. Manchester: Manchester University Press, 194–214.

Priestley, J. B. 1934: *English Journey*. London: William Heinemann.

Pugh, M. 2000: *Women and the Women's Movement in Britain 1914–99*. Basingstoke: Macmillan.

Riddell, P. 1985: *The Thatcher Government*. Oxford: Basil Blackwell.

Ryan, D. 1995: The *Daily Mail* Ideal Home Exhibition and suburban modernity, 1908–51. Unpublished Ph.D. thesis, University of East London, Department of Cultural Studies.

Ryan, D. 1999: Staging the imperial city: the Pageant of London, 1911. In F. Driver and D. Gilbert (eds), *Imperial Cities: Landscape Display and Identity*. Manchester: Manchester University Press, 117–36.

Samuel, R. 1998: *Theatres of Memory*, vol. 2: *Island Stories: Unravelling Britain*. London: Verso.

Sawyer, M. 1999: *Park and Ride: Adventures in Suburbia*. London: Little, Brown & Co.

Seyd, P., and P. Whiteley 1996: Conservative grassroots: an overview. In S. Ludlam and M. J. Smith (eds), *Contemporary British Conservatism*. Basingstoke: Macmillan.

Sharp, T. 1936: The English tradition in the town. III: universal suburbia. *Architectural Review*, 79, 115–20.

Silverstone, R. (ed.) 1997: *Visions of Suburbia*. London: Routledge.

Webster, R. (ed.) 2000: *Expanding Suburbia: Reviewing Suburban Narratives* Oxford: Berghahn.

Williams, P. 1992: Housing. In P. Cloke (ed.), *Policy and Change in Thatcher's Britain*. Oxford: Pergamon, 159–98.

Woman's Own 1987: AIDS, education and the year 2000. Interview with Margaret Thatcher. 3 Oct., 8–10.

Chapter Eleven

Nation, Empire and Cosmopolis: Ireland and the Break with Britain

Gerry Kearns

Introduction

Citizenship, affections and loyalties attach to a variety of scales from the local to the global. Culture and politics must embrace a geopolitical vision (Kearns 2003). The scaling of identities is contestable and unstable. It is unstable both because the content of the identity can be challenged (what does it mean to be British?) and because the spatial units to which loyalty is owed may exert contradictory pulls (when is an Irish person not a European?). In all spheres claimed as British, the twentieth century opened to noisy debate and struggle over precisely these questions, and particularly in Ireland. Irish nationalism posed a direct challenge to the integrity of the two spatial units to which British loyalties were addressed: the United Kingdom of Great Britain and Ireland, and the British Empire of Great Britain and its colonies. Ireland was in some respects a province and in others a colony. This ambivalence inflected Irish nationalism in so far as it was based on the rejection of British domination (Howe 2000; Kearns 2002). The British effect also registered in the ambivalent content of Irish nationalism. Part of the rejection of British rule was the refusal of Irish nationalists to see themselves as the British saw them (Deane 1997). In reaction to British views of the Irish as at times feminine and at others as uncouth, this is responsible for the hyper-masculine dimensions of some and the bourgeois dimensions of other forms of Irish nationalism (Lloyd 1993). In addition to these difficulties of scale and content bearing the impress of its British referent, Irish nationalism had other, more local, sources of instability. Again, these attach both to scale and to content.

Nationalism presents itself as wedded to the spatial unit containing its people. In the case of a diasporic people such as the Irish this reduces to a focus upon its place of origin. Nationalists often reject certain identities as too broad (international, or lacking national distinctiveness) or too narrow (parochial, or too distinctive). This scalar discipline can be seen clearly in early twentieth-century Irish nationalist attacks on various ideologies as cosmopolitan. This chapter begins by reviewing this set of arguments. The national distinctiveness appealed to is in fact a normative claim upon the future: this is how the nation should conduct itself. These goals are various and give rise to conflicting nationalisms. The second part of the chapter describes debates between competing nationalisms. In this contest it seems that nationalism's spatial discipline quickly collapses. The chapter ends by reviewing this instability and argues that nationalisms typically refer beyond the nation to more global interests in justifying their goals, and refer below the nation to communities of experienced solidarity as anticipations of the brave new national tomorrow. People are told that the national future has local portents.

Cosmopolitans

There is a version of Enlightenment thought that sees a conflict between local allegiances and the fairness of universal values. Kant described this generalized position as 'cosmopolitan' (1963). Conservatives such as Carlyle and Spengler have abused cosmopolitanism for representing the contamination of national purity by foreign ideas. Others now see cosmopolitanism as the cultural logic of globalization (Brennan 1997). These readings of cosmopolitanism were all evident in early twentieth-century Ireland.

When England invaded Ireland and tried to extirpate the Catholic religion, it forged, in reaction, an alliance between nationalism and Catholicism. In this way, Ireland as a whole, argued John Eglinton, had been denied the benefits of the Renaissance and the Reformation, leaving clerics there to abhor the 'enfranchisement of the human intellect' (1904: 13). This had left Ireland in thrall to what the socialist Frederick Ryan termed 'effete theological dogmas' (1906a: 7; Eagleton 1998). Nationalist imperatives had forced Irish thinkers to avoid matters, such as religion and the Enlightenment, that might divide the Irish people. The cosmopolitan claim was that, instead of free thought, Irish nationalists turned to literature where they could 'promote an artificial and sentimental unity in Irish life by carefully ignoring all those matters as to which Irishmen as thinking and unthinking beings hold diverse opinions' (Eglinton and Ryan 1904: 2). But these Enlightenment questions could only be postponed for 'when the hurricanes of national and racial antagonisms die away we must always come back to equity, to utility, and to righteousness' (Ryan 1905: 279).

Writing as John Eglinton (1906: 11–12), W. K. Magee claimed that:

> The day of nations...is passing away. The day of those institutions for which we have at present no more high-sounding names than 'local self-government,' 'municipal trade,' and the like, is already in progress – the period which from the point of view of the consummations of the future will perhaps appear as the democratic middle ages.

Eglinton argued that the injunction to produce works that served nationalism forced Irish writers to censor their individuality in the service of national unity. Eglinton refused the guidance of the famous nationalists of the 1840s Young Ireland movement. Adverting to their intellectual leader, he called for the de-Davisation of Irish literature (Kearns 2001a). Instead, writers should simply deliver themselves of their most profound thoughts and deepest feelings: 'literature must be as free as the elements; if that is to be cosmopolitan, it must be cosmopolitan' (Eglinton 1906: 42). Whatever interfered with the pursuit of universal goals and global solidarity belonged to the Dark Ages. The journalist Stephen McKenna argued that global citizens had to divest themselves of the legacy of the Tower of Babel. Everyone should learn Esperanto as a second language (McKenna 1903). In contrast, nationalists wanted English-speaking Irish people to learn Irish as a second language.

The most explicit cosmopolitans in politics were probably the socialists. From 1864, Karl Marx and Friedrich Engels worked to inspire and direct a global proletarian movement since known as the First International. Their *Communist Manifesto* taught that workers' primary allegiance should be to their class, whatever temporary and tactical alliances that may involve. In 1885 William Morris warned Irish workers that, while national struggles may be a short-term necessity, were they to divert attention from the goal of global insurrection they were reactionary. Instead, the Irish should 'make up their minds that even if they have to wait for it their revolution shall be part of the great international movement' (Lane 1997: 116). Nationalism would never be enough and might not even be progressive. An Irish member of Morris's Socialist League argued that 'the tendency of the age was towards internationalism not nationalism. It was absurd to think that the separation of Ireland from England would alone benefit the working men of Ireland' (Lane 1997: 121).

Pure cosmopolitanism was rare, but it was used as a stick for beating people and views judged to be un-Irish. Douglas Hyde wrote of 'the effacing hand of cosmopolitanism' (1986a[1886]: 76). Global identities erased local distinction and thus: 'cosmopolitanism may be an attractive idea, but it can scarcely appeal to those who see all that makes life worth living, stamped out in its name' (Coffey 1911: 283). Cosmopolitanism threw out spiritualism and welcomed materialism in its stead:

A soulless cosmopolitanism never yet achieved anything of permanent value, and the Gaelic Movement is in its essence a vigorous protest against the 'confounding of differences,' an organised reaction against the rather prevalent conception of civilisation, which would seem to have for its ideal the doing away of the Curse of Babel to the great advantage of those who buy and sell. (Perceval 1905: 366)

Cosmopolitanism would usher in a pagan modernity that would displace the Catholic faith of true Ireland:

In Ireland our ideals of nationhood are naturally bound up closely with those of our faith, let us *there* be in touch with the great world, and not through a cosmopolitanism that can develop logically into nothing but a suicidal anarchy. (Gill 1907: 187)

Maurice Joy convicted William Butler Yeats of a 'vagabond intellectualism', producing literature that was not part of the 'authentic business of life' because it had neither 'reverence' nor 'respect' for 'the intense inner life of the people' (1905: 260). Refusing to root itself in the attitudes of his Irish contemporaries, Yeats's art was hardly national at all. Cosmopolitanism as rootless, materialistic and alien was the avatar of modernity. At a time when Ireland was 'far gone in modernity', the Celtic spirit had risen again and had 'marshalled our emotions and guided our desires' towards 'transcendental things' (Goddard 1903: 358–9). In 1903 Douglas Hyde and Maud Gonne left Yeats's Irish National Theatre Society because it was unreliable as a vehicle of nationalist propaganda. Commenting on the split, Arthur Griffith attacked such as Yeats for accepting foreign models such as Henrik Ibsen for Irish drama: 'cosmopolitanism never produced a great artist nor a good man yet and never will' (Foster 1991: 297).

The Instability of Content in Nationalisms

Nationalists, then, argued that cosmopolitanism was insufficiently distinctive to suit Irish needs. However, the nation referred not only to a space but also to a model of the sort of society that should occupy that space. Nationalism was an ideology in the service of distinctive projects. Cosmopolitanism was rejected in the name of quite disparate social visions. In the final part of this chapter, I will argue that those visions were unable to avoid making an appeal to precisely the sorts of universals for which cosmopolitanism was berated. For now, I want to recover some of this debate between nationalisms by outlining just seven of those that Irish people invoked.

The historic nation

All nationalists agreed that Irish people had a historic right to nationhood, even if they disagreed about what that nationhood should look like. History served to reject British calumnies about the Irish, as the veteran nationalist Charles Gavan Duffy explained when urging that the Irish must set aside British books: 'Ireland and Irishmen suffer wrong from systematic misrepresentation' (1894: 39). For Perceval history could be 'a stimulus to rehabilitation or natural development from within' (1905: 363). Ireland had a glorious past, having been the beacon of Christian scholarship during the so-called Dark Ages. Unlike that of the British, the Irish record was, argued Hyde, 'unstained by oppression of any men, untarnished by avarice of anything, and undimmed by murder' (1986b: 180). Ireland had its heroic period of warriors, queens, nobility and comradeship. Preserved in songs, myths and fables, these connected living Irish people to a time when gods walked the earth. The English occupation lessened but did not stanch the flow of inspiration from Ireland past to Ireland present. Ireland's past could serve as immunity against British materialism. Ireland alone could be virtuous; colonialism drove the Irish away from their better, tolerant selves: 'opinion is suppressed in Ireland, because we are, as it were, under Martial Law' (Synan 1905a: 271). Irish freedom was on every hand contradicted by British colonialism.

These historical themes were hegemonic in Irish nationalism. Of course, in Presbyterian Ulster and among Protestants in all parts of Ireland, there were some who shared aspects of this heritage but drew different conclusions about the British connection. Even among republicans, a few warned about this view of history. The Quaker Alfred Webb thought that the Irish dwelled too much in bygone days: 'is it not best to look hopefully forward, garnering the experiences of the ages and of all peoples, rather than to seek to live in and by the far past of our own country alone?' (1904: 144). Conversely Arthur Synan insisted that Irish people were not in fact living in the past but were dealing with the current legacy of past injustice: 'the reason why historical happenings have such a powerful effect on Irishmen, is that those happenings still produce their results; the wrongs of history are still unredressed' (1905b: 67). Other nationalists focused almost exclusively on past rather than present discontents and followed John Mitchel in seeing the undoing of British conquest as the only thing needful (Kearns 2001b). This Manichaean vision frequently heightened nationalist rhetoric and even rendered it abusive, and some, such as the co-operative campaigner Æ, were disappointed: 'it is surely time to get rid of that blind spirit which refuses to see sincerity outside its own beliefs, which confines all the virtues within the narrow boundary of its own expansion, its own creed, its own party' (1904a: 131). The historical and Manichaean strain in nation-alism made this unlikely.

The political nation

Nationalists agreed that Ireland should be a political nation once again. Independence was the necessary prelude to the training of a people in citizenship: 'just as the noblest men find their highest satisfaction in self-knowledge and self-discipline, so nations will come to find their true ideal in developing their own mental and moral wealth' (Ryan 1904: 117). Since the 1840s, Irish nationalism had been split over tactics: was the surest route to independence through constitutional agitation or violent insurrection? Many argued that the British only made concessions when prodded by pike or harried by musket: 'if we look into Irish history we shall find that every important measure of Irish reform . . . was the result of a violent exhibition of hostility to the Government' (Synan 1905b: 69). While physical-force nationalism worried the property-owning classes in Ireland, it was never-theless true that Ireland's most effective moral-force leaders, including Daniel O'Connell and Charles Stewart Parnell, had presented themselves to the British as the best alternative to the anarchy of outright rebellion. There were times when moral and physical force appeared to work in tandem, others when one predominated and the other regrouped, others still when Irish people sang to the tune of other nationalisms, literary or linguistic, as in the early twentieth century. Some worried about this oscillation:

> It is impossible that a language movement, or an art movement, or a manu-facture movement can ever take the place of a political movement. As long as questions of peace and war, of land and labour, of housing, of taxation, of provision for old age . . . exist, a political movement of some kind there must and will be, however apathetic things might become for a season. (Webb 1904: 142)

There was little agreement over what form independence should take, with limited discussion over forms of government in the abstract. Hutchinson saw democracy as 'class tyranny' (1906: 262) in the workers' interest. Frederick Ryan (1906b) replied that politics had hitherto been the preserve of a middle class who had served only themselves. In the main, however, forms of government were debated in more pragmatic terms. In 1905 Arthur Griffith founded Sinn Féin, a party committed to a policy of self-reliance for Ireland. The Irish should not stoop to asking the British Parliament for their freedom. Instead, Sinn Féin policy was to begin self-government in Ireland with political representatives already chosen by the

Irish people. Members of Parliament elected for Irish constituencies should withdraw from Westminster and join with delegates from each Irish local government unit to form a National Council to run the country: 'with our representatives assembled in Dublin, the eyes of the Irish people would naturally be turned on Ireland, and especially upon their National Council assembled in Dublin and composed of their own elected representatives' (Sinn Féin 1915: 2). Taking up an earlier suggestion of Daniel O'Connell, this council was charged to appoint local magistrates, or arbitration courts, so that Irish people might settle their differences in front of Irish and not English authorities. Something like these arrangements was established in defiance of British rule in 1919 with the creation of the first Dáil, or Irish parliament. These parallel institutions might make Ireland ungovernable and prepare the way for formal separation from Britain.

There was extensive debate, over tactics rather than principle, about how separate Ireland should be from Britain. Most committed nationalists were in favour of a republic, wanting no connection with the British crown. However, in deference to local Unionists and British intransigence many would consider something less. Griffith (1904) wanted Sinn Féin to be an all-inclusive movement, and advocated a dual monarchy, after the example of Austria-Hungary. Ireland would be a separate state with a head of state called the king, or queen, of Ireland. This could be the same individual as the British head of state, but they would not sit on the Irish throne as the British monarch. Griffith hoped this might be enough to convert cultural-nationalist Unionists into political-nationalists; in general it was not. In 1926–7, and thus after Griffith's death (1922), the proposal was revived by Kevin O'Higgins in a vain last effort to avoid partition (de Vere White 1966). A second alternative, suggested by those more sensitive to British rather than Irish objections to complete separation, was self-governing dominion, recognizing the British monarch as the head of the empire rather than head of state. Dennehy believed that 'we cannot hope to accomplish anything of a practical kind for Ireland, by availing ourselves of every opportunity for manifesting our hostility to the British Empire' (1905: 13), but the empire was anathema to many: 'our patriotism, our love of country, rests on Ireland alone, and not on the British Empire' (Sweetman 1905: 199). This was certainly the view of the secret societies plotting revolution, but the main party of insurrection, the Irish Republican Brotherhood, exercised little influence until Tom Clarke returned from the USA in 1907 to rebuild it. Thereafter, its members joined all nationalist organizations to promote republicanism through rebellion. By 1915 they dominated the linguistic nationalist organization, the Gaelic League, and its president, Hyde, resigned in their favour. In 1917 Griffith, president of Sinn Féin, did likewise. The anti-Treaty movement after 1921 was staffed by these physical-force republicans.

The linguistic nation

When political nationalism waned, linguistic nationalism waxed, seeming to offer a haven from political strife. Speaking of the cultural nationalism of the language movement and the Irish Literary Theatre, Hyde told its supporters that 'by national, he meant something absolutely uncontentious, non-political and non-sectarian (hear, hear)' (*Freeman's Journal* 1900: 6c). Over the nineteenth century, scholarly interest in Gaelic literature developed apace. At the same time, the language as spoken was in a tailspin dive. If, early in the century, every other inhabitant could speak Irish, by its end few more than one in ten did. Even fewer could read and write in Irish. Many among the polite classes would have agreed with Robert Atkinson, a contemporary scholar of the Irish language, that 'all folk-lore is at bottom abominable' (Dunleavy and Dunleavy 1991: 210). There appeared to be an ocean between the speech of the illiterate peasant and the poetry of the ancient sagas. Onto these waters Douglas Hyde set sail with his lecture of 1892 on 'The necessity for de-Anglicising Ireland'. The Irish, he suggested, had always refused to assimilate to the culture of the colonizer. As Fr. O'Hannay explained, Anglicization had 'resulted not in the making of Englishmen but in the unmaking of Irishmen' (1906: 839). Such nondescripts might not deserve independence: 'in Anglicising ourselves wholesale we have thrown away with a light heart the best claim which we have upon the world's recognition of us as a separate nationality' (Hyde 1894: 119). The Irish language, habits, customs and folklore were 'the bricks of nationality' (1894: 129). Art critic George Moore believed that 'it is through language that a tradition of thought is preserved, and so it may be said that the language was the soul of a race' (1901: 47). Irish was the most coherent example of a Celtic civilization which almost everywhere else in Europe had been trampled down by Roman legions. On these lines the Irish might rebuild their nation to become again 'one of the most original, artistic, literary, and charming peoples of Europe' (1901: 161). Language could renew the Irish ethnie: 'in a word, we must strive to cultivate everything that is ... most Gaelic, most Irish, because in spite of the little admixture of Saxon blood in the north-east corner, this island *is* and will *ever* remain Celtic at the core, far more Celtic than most people imagine' (1901: 159).

In 1893 the Gaelic League was founded. Hyde was president. The aim was to reverse the decline of the Irish language and thus civilization. In areas where Irish was still known, the League sent out peripatetic teachers to encourage the young to take up the language of their grandparents. They provided Gaelic entertainments, particularly dances, in these predominantly rural areas of the west. Between 1896 and 1901, 135,000 copies of the

League's basic language primer were sold (Grote 1994: 73). It established a college for the training of teachers of Irish in each of Ireland's four provinces. By 1906 there were between 500 and 1,000 branches of the League in Ireland with somewhere between 50,000 and 100,000 members in total (Grote 1994: 77). The League lobbied successfully to get Irish taught as a language for examination in schools, and campaigned for Irish to be a compulsory subject for admission to the new National University (1908), eventually prevailing. The League organized a torchlight procession through Dublin of between 100,000 and half a million people (Grote 1994: 97).

Linguistic nationalism forged a demotic nation. It countered the snobbery that saw peasant Ireland as but superstitious and primitive. Only the living Irish of the peasant could recover the place names, the songs and the vocabulary of pre-conquest Ireland: 'the language of the western Gael is the language best suited to his surroundings' (Hyde 1986a: 77). The peasant was the authority who could unlock the meaning of those sagas and poems in the manuscripts venerated by antiquarians and scholars. Irish literature, peasantry and nation would flourish together. Patrick Pearse said that, through its celebration of Irish literature, Hyde's 1899 *A Literary History of Ireland* was 'equal in effect of *Uncle Tom's Cabin*' (Dunleavy and Dunleavy 1991: 208). This linguistic nationalism redrew the imaginary geography of Irish identity. In the first place, it rejected provincialism. From the ancient sagas some constructed a genealogy of the Irish language, tracing it back to the masons who built the Tower of Babel (Tymoczko 1994), with Seaton insisting on the purity and independence of Gaelic as 'one of the first languages spoken on earth' (1898: 223). Thomas MacDonagh, the lecturer in English at University College Dublin executed as one of the leaders of the 1916 rising, realized the implications of this geography. The renewal of Irish literature must come, he thought, from Irish-speaking playwrights in the rural west, from those steeped in 'Gaelic ways of thought and life' (MacDonagh 1920: 158).

Many took their first steps towards political, even insurrectionary, nationalism through an Irish-language class: 'in 1906 when I joined the Irish Republican Brotherhood, on the invitation of Sean O'Casey, the great majority of its members in Dublin, which was incomparably its strongest centre, had come in through the Gaelic League' (De Blaghde 1972: 34). The 1916 rising was led by people who, in the main, had been shaped by the Gaelic League. The League was promptly proscribed by the British government. Yet some found the language movement both mystical and exclusive. To define a nation by language, as did Hyde, although by birth an English-speaking Protestant, was unnecessarily restrictive: 'to make Irish, or even the desire to acquire it, the test of Nationalism, would shut out some of the best men who have served the cause of Irish liberty in the past' (Ryan 1904: 217). Indeed, in direct contradiction of the claims of Hyde, Eglinton (1906: 7) insisted on the central importance of the Anglo-Irish tradition: ' "Irish" language is indeed only a title of courtesy: the ancient language of the Celt is no longer the language of Irish nationality'. This Anglo-Irish heritage animated literary nationalism instead.

The literary nation

Literary nationalism was, in the main, an English-speaking cultural movement that, like linguistic nationalism, marked the disaffection of many with political nationalism. In 1891 Yeats was central to the founding of the Irish Literary Society in London, and in 1892 of the Irish National Literary Society in Dublin. Brown (1999) suggests that Yeats thought the hiatus in political agitation created the space for his poetry to shape anew the Irish identity. Yeats asked: 'may not we men of the pen hope to move some Irish hearts and make them beat true to manhood and to Ireland?' (1964a: 19). Poetry did not, however, garner the audience Yeats wished because, he thought, the people did not read much outside cheap journalism, 'and so from the very start we felt that we must have a theatre of our own' (1999a: 410). With the financial and moral encouragement of Augusta Gregory, Yeats founded the Irish Literary Theatre in 1899. The aim was to present Irish plays in Dublin countering the stereotypical representations of Irish men and women that so amused London audiences. Rejecting 'the stupefying memory of the theatre of commerce', the Irish Literary Theatre would mount plays 'for the most part remote, spiritual, and ideal' (Yeats 1961a: 166). These plays were about Irish matters. Drama should engage with ideals, and that meant taking up local enthusiasms. In Ireland, Yeats felt, artists 'will find two passions ready to their hands', nationalism and the supernatural (Yeats 1961b: 204). Yeats associated himself with the Protestant aristocracy of Ireland, and his exploration of symbols and ideals to forge and direct an Irish nation contested the monopoly claimed over Irish hearts and minds by the Roman Catholic Church. Hutchinson (1987) insists that intellectuals such as Yeats turned to Gaelic symbols precisely in order to take authority away from religion.

Many of the plays were historical. Gregory was sure that 'to have a real success and to come into the life of the country, one must touch a real and eternal emotion, and history comes only next to religion in our country' (1914: 19). Historical plays could show heroic characters and underline the virtues that made them venerable. In Gregory's *Kincora*, Brian Boru is celebrated not for his martial valour but because he made war in order to secure peace: 'this tossed, tormented country has to be put in order... and travel whatever road God laid out for it, without arguing and backbiting and the quarrelling of cranky bigoted men' (Gregory 1983: 57). To some extent, Yeats moved further back in history to circumvent contemporary schisms of class and religion. The plays of the literary revival were in English, although they tried to convey an Irish dialect. The playwright John Millington Synge was sure that 'within the present generation the linguistic atmosphere of Ireland has become definitely English enough, for the first time, to allow work to be done in English that is perfectly Irish in its essence' (1982a: 384). Yeats' *Kathleen Ní Houlihan* was an early successful use of dialect on stage, with dialogue rewritten by Lady Gregory and rendered in the form of the dialect spoken by country folk near her family home. Synge developed the technique and used it for plays set in the present, plays explicitly confronting the romantic idealization of the peasantry found in many nationalist works. To look with honesty upon the condition of Ireland required such close attention to time and place that historical plays were 'relatively worthless' (Synge 1982b: 350). Synge showed the evil that could follow from romantic, nationalist illusions and braggadocio. He showed as cramped any moral imagination that romanticized poverty. Yeats believed that, in drawing upon the speech patterns of the Aran Islanders, Synge found a concrete form of expression immune to nationalist bombast, a language that 'could not express, so little abstract it is and so rammed with life, those worn generalisations of National propaganda' (Yeats 1961c: 335).

The tensions created by this attempt to forge an Irish conscience through drama are evident in responses to Yeats's *Countess Cathleen* (1899) and Synge's *Playboy of the Western World* (1907). Cathleen is a Protestant, a landowner, who sells her soul to spare her tenants from famine. The plot served Yeats's fantasy that in an independent Ireland the Anglo-Irish aristocracy might retain its social and economic privileges. However, the play was read an insult by many Catholics and nationalists. Frank Hugh O'Donnell condemned the play for the blasphemy of suggesting God could ever condone the selling of souls and, sight unseen, Cardinal Michael Logue agreed (Brown 1999). Watson (1994) is surely right to remark that the workers of Dublin did not like being reminded of the rural poverty from which so many of them were removed by no more than a generation. As Howes (1996) remarks, in the folk memory of the poor it was the peasants who were asked to sell their souls by converting to Protestantism in return for soup. The attacks on the play were marshalled along exactly the lines of class and religion that its plot had sought to suspend. The *Playboy* riots had similar roots. In 1921 in 'The Leaders of the Crowd' Yeats despaired of the chance of breeding tolerance in Ireland: 'They must to keep their certainty accuse | All that are different of a base intent' (Albright 1990: 232). In an unpublished essay Synge was more direct: 'the Gaelic League is founded on a doctrine that is made up of ignorance, fraud and hypocrisy' (1982c:

399). The League pretended that an Irish-speaking Ireland was a real possibility and that Ireland rested upon a spiritual and pure peasantry untainted by material or carnal desires. The socialist journalist Robert Lynd (1905: 376) warned of the censorship to come: 'if we begin by cramping our artists...we shall only succeed...in mutilating their individuality to such a degree that we shall never possess any great art'. This was a price that those who saw Ireland as a religious nation might well have been willing to pay.

The religious nation

To a fair extent, the unity of Ireland was forged in the fire of English invasion and tempered in that of the colonial administration that followed. The justification for the seventeenth-century invasions was given as religious and the land transfer that followed replaced Catholic with Protestant landowners. By 1700 four-fifths of land the island's land was owned by the one-fifth of the population who were Protestant. Various civil disabilities imposed upon Catholics maintained this settlement. Over the nineteenth century many of these disabilities were set aside, at least for the middle-class Catholic. Nevertheless, it remained impossible to think of Ireland outside the religious genealogy bequeathed by the Reformation.

The relations between the Irish Catholic Church and nationalism were always close, if fraught. The Church was in many ways a conservative, even reactionary institution. In 1848 Italian nationalists had challenged the temporal power of the Pope and ejected him from Rome (Duffy 1997). Ever after, physical revolutionary nationalism was anathema to the Pope. From the founding of the Irish Revolutionary Brotherhood in 1858, there was an underground movement of secret societies, their oaths placing loyalty to the cause above individual conscience, and incompatible with the moral leadership claimed by the Church. The leading Irish bishop of the third quarter of the nineteenth century, Paul Cullen, 'carried most of his episcopal colleagues with him in a policy of denying the sacraments to the Fenian revolutionaries of the period' (Miller 1973: 5). At the local level, not all parish priests followed Cullen's lead. Many nationalists tried to reassure bishops and Pope that there was no conflict between the claims of God and country. The Home Ruler John Dillon was direct: 'the man who is a good Catholic is a good Nationalist' (Larkin 1979: 174). A decade later the matter was even clearer for the journalist David Moran: 'if a non-Catholic Nationalist Irishman does not like to live in a Catholic [nation] let him... give up all pretence to being an Irish Nationalist' (Miller 1973: 41–2). For Moran, the purpose of nationalism was to re-establish a Catholic nation that would stand against the materialism and modernism of the rest of Europe.

The cause of the religious nation rested upon the identification of the Irish people with the Catholic faith and of the Catholic faith with national values. In the history textbooks used by the Christian Brothers: 'the emphasis is on what Irish Catholics have suffered for their faith, rather than on what Irish people have suffered under a colonial rule' (Coldrey 1988: 121). To restore those happy pre-Reformation days required not just spiritual but also political action. The Church could thus see a use for nationalism. In any case, it had to accommodate itself to nationalism, for such was the settled wish of so many of its constituents. In 1898 the archbishop of Tuam explained why he could not anathematize the United Irish League: 'the people have taken up the matter so warmly as to have gone beyond the control of the clergy' (Miller 1973: 20).

Peter van der Veer (1994) has shown that the contradiction between the nationalist reverence for history and the religious emphasis upon timeless values is only ever apparent. Nationalism seeks its legitimacy in a historical compact with a pure past yet it inevitably reinterprets that past to serve modern projects. Religious interests use the state to secure various monopolies that they desire (for example, over education) or to remove the indignity of competing moral frameworks from the public sphere (through various forms of prohibition). This was the prize of a Catholic state for a Catholic people. Nationalism, however, can also make use of religion.

The rebellion of Easter week 1916 was steeped in religious imagery. Many of the rebels were devout Catholics. Pearse was exceptional in his identification with the person of the crucified Christ, writing to his mother from prison that: 'you too will be blessed because you were my mother' (Ó Buachalla 1980: 377). Soon after, he was shot. Pearse was one of seven ex-Christian Brother pupils among the 14 rebels executed in Dublin. Although the Church initially condemned the rising as irresponsible and socialistic, the executions and the stories of the piety of the rebels quickly produced a change in attitude: 'it is said that during the rebels' occupation of strategic buildings during Easter Week, the Rosary was recited on the hour, every hour' (Kenny 1997: 62). As martyrs for Ireland, the rebels easily passed from the secular to the sacred: 'since political demonstrations in Dublin were forbidden in the weeks following the Rising, memorial masses for the fallen and executed rebels became a focus for the growing public sympathy' (Miller 1973: 341–2). This communion allowed nationalists to draw upon the rhetoric of religious nationalism to consecrate their land and their heroes and call for divine protection against the English.

The power of this consecration gave the Church an opportunity to reclaim the rebellion as an act of religious nationalism, which it did. As such, it was harder for opponents or even critics of religious nationalists to get a hearing. These included, of course, Protestants such as Yeats. They also included Æ (1904b: 47), who convicted Christianity of being loveless:

'if we drink in the beauty of the night or the mountains, it is deemed to be praise of the maker, but if we show an equal adoration of the beauty of man or woman it is dangerous, it is almost wicked.' These sanctions on sexual expression were central to religious nationalism. Nationalism 'appropriates the disciplinary practices, connected to the theme of the management of desire, in the service of its own political project' (van der Veer 1994: 201). Finally, the anti-materialism of Catholic social thought was soon too easily used to excuse independent Ireland's poor economic performance. It seemed that Horace Plunkett might have been correct in 1904 with his indictment of Catholicism as a block to the pursuit of material well-being (Kenny 1997). It was also used to sweeten the shelving of the social questions that economic nationalists had been urging.

The economic nation

Economic autarky was important to Arthur Griffith's Sinn Féin policy. Only with independence could Ireland begin to make use of its resources to serve its own people. Irish people paid more in taxes than they received in benefits. It was claimed that British taxation and excise policies had distorted the Irish economy so that, at critical moments, Irish industry and agriculture could not compete with their English counterparts. Comparative advantage once secured, free trade ensured that the Irish could never catch up. Fiscal autonomy was of primary importance to Griffith when he went to London in 1921 to negotiate the treaty to end the war with Britain (Pakenham 1972). This did not satisfy those, such as de Valera, who were most concerned about the fate of the political nation and for whom the question of an oath of allegiance to the British monarch was non-negotiable. Upon this rock the ship of state ran aground and civil war ensued. Different ideas about what the nation is for can coexist until hard choices seem unavoidable.

The most resolutely economic reading of Irish nationalism in the early twentieth century was that of James Connolly. He accepted the political but not the economic programme of Arthur Griffith. With Sinn Féin's 'economic teaching as expounded by my friend Mr Arthur Griffith in his adoption of the doctrines of Frederick List, Socialists have no sympathy, as it appeals only to those who measure a nation's prosperity by the volume of wealth produced in a country, instead of by the distribution of that wealth amongst the inhabitants' (Connolly 1909a). Connolly was a socialist. He believed in the state ownership of the means of production. Furthermore, he believed that nationalism in Ireland was the most direct route to achieving socialism. Socialism required control of the economy in the name of the people; thus in Ireland English rule had to end: 'thus inspired by another ideal... the Socialist Republican Party of Ireland arrives at the same conclusion as the most irreconcilable Nationalist' (Connolly 1897a). The proletariat was the most thoroughgoing nationalist class, having most to gain from ending ties with Britain, and what it had to gain was a Socialist Republic: 'Nationalism without Socialism – without a reorganisation of society on the basis of a broader and more developed form of that common property which underlay the social structure of Ancient Erin – is only national recreancy' (Connolly 1897b). The reference to Ancient Erin attempts to ground socialism in explicitly Irish traditions and to avoid the disabling charge of cosmopolitanism. The Eden from which religious nationalists had been expelled was Catholic; Connolly's was one of primitive communism.

Connolly tried hard to avoid conflict with the Catholic Church, arguing that socialism and Christianity related to different spheres: 'we know that Socialism bears upon daily life in the workshop, and that religion does not' (Connolly 1909b). In reply to the claim that socialism would destroy religion, Connolly (1908) was reassurance itself: 'when socialism wins, the Catholic church will find ways to accommodate to it, as it does to any established order.' He aimed at 'presenting Socialism so that it will appeal to the peculiar hereditary instincts and character of the people among who [he was] operating' and in Ireland this meant 'getting entangled in the question of religion' (Connolly 1914). He would restate socialist principles in Catholic terms: 'to the average non-Socialist Irishman the idea of belonging to an international political party is unthinkable...But he belongs to a church – the Roman Catholic Church – which is the most international institution in existence' (Levenson 1973: 148–9). He was keen to use national symbols: 'we honour St Patrick's day (and its allied legend of the shamrock) because in it we see the spiritual conception of the separate identity of the Irish race – an ideal of unity in diversity, of diversity not conflicting with unity' (Connolly 1916). A week before the still secret rising, Connolly told the armed trade unionists of the Irish Citizens' Army that there would definitely be a fight, but that: 'in the event of victory, hold on to your rifles, as those with whom we are fighting may stop before our goal is reached. We are out for economic as well as political liberty' (Levenson 1973: 292).

The ethnic nation

In view of its importance in academic theories of nationalism, strict ethnicity, defined in terms of biological descent, appears scarcely at all in debates over Irish nationalism in Ireland. Genealogy played a most important role among the Irish of the diaspora, with the avowedly Catholic Ancient Order of Hibernians subject to schism between 1884 and 1905 over whether it would admit people of Irish descent alongside those of Irish birth (Miller 1973). In most cases, however, nationalists stressed other matters. Thus Connolly, speaking against the idea that his nationalism was divisive by being based on birthright, explained: 'I do not care where you were born – (we have had Jews, Russians, Germans, Lithuanians, Scotsmen and Englishmen in the [Socialist Party of Ireland]) – but I do care where you are earning your living, and I hold that every class-conscious worker should work for the freedom of the country in which he lives, if he desires to hasten the political power of his class in that country' (1911). In 1899 one Protestant civil servant wrote to Eoin MacNeill, then prominent in the Gaelic League: 'as to race, half the Leinster Catholics have Saxon blood, and as to religion we are all just what we were born – these things are really *accidents* – and the true test is what is a man's attitude to Ireland' (Miller 1973: 40). MacNeill himself emphasized tradition over biology as a basis for identity: 'there is no existing . . . Anglo-Saxon race, and no Celtic race. Each of the groups to whom these names are popularly applied is a mixture of various races which can be distinguished' (1919: 2). Perhaps the racism of British imperialist attitudes towards the Irish made arguments about characteristics carried in the blood somewhat offensive. Furthermore, with Irish genealogy so difficult to forget, the amnesia that allowed the English to believe themselves to be of untainted bloodstock was perhaps harder to achieve. Certainly, when the term Irish race is used, it is evoked in most cases as a cultural rather than as a biological entity. To a limited extent, the late writings of Yeats posit a biological basis to society and defend eugenics, but here it is in defence of class, not national survival.

The Instability of Scale in Nationalisms

Nationalism, then, was a rescaling of political identities. Each form of nationalism was vulnerable to the charge that it did not capture the essence of the people of the territorial unit corresponding to the nation. No one questioned but that the Irish people had historical, political, linguistic, literary, religious, economic and ethnic dimensions. Prioritizing one in particular among these dimensions, nationalists of a particular stripe could devalue others as not sufficiently Irish. A strong break is asserted between the national and the supra-national levels when it comes to identities. No such problematic break below the national level is recognized at this date. There is very little evidence of attempts to dismiss various nationality identity claims as merely parochial. Yet each of these nationality claims made use of both local and cosmopolitan dimensions. In very broad terms, we might say that the local level was invoked in order to ground identities in affective communities. Solidarities seem to have rested upon or at least made reference to face-to-face communities of trust.

The historic nation was thought to rest upon an ancient tribal society. It was also heir to a story of invasions and settlements rendering the question 'Where yet was Ireland?' dependent upon a perhaps arbitrary answer to the earlier question 'When was Ireland?' The political nation was dispersed among parishes and constituencies. This was where much politics happened. The secret societies in particular had a decentralized structure of cells, with 'centres' reporting to higher levels. Although they differ in their explanations for the patterns they observe, both David Fitzpatrick (1978) and Tom Garvin (1981) emphasize the profoundly regional nature of political nationalism. Garvin shows that political nationalism drew upon, and at times competed with, evolving but long-standing local traditions of collective action, and that these traditions were stronger in some regions, such as Connacht and Ulster, than in others, such as Munster. The linguistic nation relied upon local reading groups and peripatetic teachers. In this case, there was the further complication of dialect. Standardized grammars and vocabularies were needed for teaching purposes, yet this would be to elevate one regional dialect over the others. There was quite an unseemly struggle between Ulster and Munster in this regard. Pearse was particularly anxious that language primers should not be based exclusively upon Ulster pronunciation.

The literary nation, at least in the hands of Yeats and his co-workers, found Irish national values in the everyday life and cultural traditions of one particular region, the west (Gibbons 1996). As Mulligan notes, the west was treated as 'a cultural reservoir from which the long forgotten and neglected gems of nationality can now be extracted' (1996: 121). This was the region to which nationalists went to lose themselves in a contemporary past, and which they invoked as ballast to stabilize the ship of state else listing in the waters of European modernity. Religious nationalism was grounded not only in the everyday and weekly practices of parish life and the spatialized rituals of pilgrimage but also upon the regional autonomy of bishops and archbishops. And then there was the religious geography of the island, always threatening to make confession a regional rather than a national matter.

The economic nation, again, was rooted in a geography of sectoral specialization. Ulster was the most obvious anomaly when it came to discussing the benefits of economic autarky. Protectionism worried Ulster industrialists, fearful of losing their British markets. Connolly was anxious to show the workers of Belfast that their primary grievance was with the local capitalists who sweated their labour. They must throw off the 'yoke of capitalism', and the nationalist movement could be their most effective and essential first step (Connolly 1917: 283). Once more, though, the communities of solidarity to which Connolly appealed, as the first glimmers of the new tomorrow, were both local and historical. Connolly argued that the class struggle humanized the workers by giving them self-respect through autonomous action (Fennell 1985: 58). Through the use of the sympathetic strike, workers might build solidarity from one local experience to another and, as Connolly promised in 1913, 'in the midst of the storm and stress of struggle [the workers' movement] solidifies into a real revolutionary force' (Anderson 1994: 20). The experience of the Irish with local co-operatives was an effective lesson in socialist values for it 'will also lead them to appreciate the advantages which might be derived from cooperation on a national scale' (Connolly 1898).

Each of these nationalisms, then, was vulnerable to the charge that it was, in reality, parochial, or regional, or, at best, provincial. Far from embarrassing these nationalisms, the local level was invoked as place where people learned to experience their distinctive Irish interests. If the sub-national was the level of experience, then the supra-national bid fair to be the level of principle. Each of these nationalisms made appeal to geographical realities beyond Ireland in justifying the programme they wished to follow at home. Nationalist discourses were not geographically self-contained. Removal of English colonialism would allow the re-emergence of an Irish nation that had, on the evidence of history, distinctive benefits to offer the rest of the world. Ireland's record as the beacon of Christian civilization during the Dark Ages promised further global benefits from its independence in the future. European and global discourses were unbalanced without the distinctive spirituality of the Irish voice. If modernity was dangerous, then, Ireland free might prove an effective bulwark. This historic nation might justify its political claims with reference to existing global discourses of freedom and of the rights of small nations. Much effort went into getting the Irish a seat at the Paris Peace Conference, where the fates of Europe's small nations were to be decided. Realpolitik ensured that the USA set the Irish to one side in order to retain British co-operation in the reorganization of the rest of Europe, but much nationalist rhetoric in the period before 1919 can only be understood as an attempt to present Ireland as the archetypical small nation. The Irish deserved freedom because they made sure they looked like the cases that Woodrow Wilson had in mind. The metric was external.

The linguistic emphasis in nationalism was certainly reinforced by this desire to fit the Wilsonian paradigm. Language, however, also drew the Irish out of Ireland in other ways too. The scholarship on the Irish language was very much European. Furthermore, the place of language as guarantor of Irish values drew many, such as James Joyce, towards a sort of linguistic geography that emphasized diffusion rather than autochthonous growth (Cullingford 2001). If the Irish language had any virtues, then, many were borrowed from elsewhere. This linguistic geography gave the Irish a direct link to a cradle of world civilizations in Phoenicia but at the expense of devaluing the purity that had, for so many, almost a racial or religious value. Closer to hand, there appeared to be good grounds to speak of a family of Celtic languages at the western edges of Europe. This gave rise to a pan-Celtic movement, but here the involvement of the Gaelic League proved contentious. When Pearse attended an Eistedfodd in Wales he was criticized for appearing to give official approval to a non-nationalist language movement.

Literary nationalism likewise often looked outside Ireland for its justification. Synge (1982d: 352) was rather dismissive of the early works of the literary revival, assuring French readers that 'it is the old [Irish] literature, and that alone, which is truly of European importance' ('c'est la vieille littérature et elle seule qui a une véritable importance européenne'). Joyce found the early work of the Irish Literary Theatre parochial because it bound itself to the prejudices of its local audience: 'a nation which never advanced so far as a miracle-play affords no literary model to the artist, and he must look abroad' (1959: 70). Yeats (1961b: 207) had come to agree, suggesting that while the materials for Irish art should be local, their development would have to be done in such a way as to please the artist him- or herself. In its art, 'no nation, since the beginning of history, has ever drawn all its life out of itself' (Yeats 1964b: 144). A national literature might unite Ireland by being a common object of veneration, but in order to do so it had to be outstanding in the eyes of the world's most demanding literary audiences, and these were found in Europe: 'I thought we might bring the [Catholic and Protestant] halves [of Ireland] together if we had a national literature that made Ireland beautiful in the memory, and yet had been freed from provincialism by an exacting criticism, a European pose' (Yeats 1999b: 105).

As Connolly reminded them, Irish Catholics were members of one of the most international organizations on the planet. Although some might chafe against Rome's involvement in Irish domestic politics, Irish Catholics sought papal approval of their struggle for independence. Through Count Plunkett, the rebels 'actually solicited the Pope's blessing' upon their Easter Rising (Miller 1973: 341). Religious nationalist discourse in Ireland referred in detail to the various encyclicals upon the involvement of Catholics in politics, seeking to justify their involvement in nationalist movements in the face of apparent papal objections. Furthermore, the goal of an independent Ireland would be to further God's holy work not only in Ireland but also beyond. A Catholic Ireland with full control over the education of its youth could supply the missionaries who might protect the Irish diaspora from proselytism and then turn to the conversion of Asia and Africa: 'the new consciousness of "Irishness" which the [cultural] revival promoted also included...a re-enactment of Ireland's illustrious missionary past' (Hogan 1990: 97).

The economic nationalism of Arthur Griffith drew upon the economic ideas of the mercantilist Friedrich List, and illustrated their value with reference to Hungary, where greater political autonomy had driven economic revival. However, Griffith's economic ideas were among the most insular of the nationalisms discussed here. On the other hand, the internationalism of socialism was asserted loudly, perhaps most frequently by its opponents. This charge was not easy to duck. Connolly's journalism was internationalist in flavour, and he made no secret of his desire to see Irish workers act in solidarity with the proletariat abroad. In response to the internationalism of capitalism, the workers must assert an internationalism of fraternity and sorority. He saw capitalism as the destroyer of national distinctiveness: 'I regard each nation as the possessor of a definite contribution to the common stock of civilisation, and I regard the capitalist class of each nation as being the logical and natural enemy of the national culture which constitutes that definite contribution' (Connolly 1915). Workers had to operate internationally to fight the power of capital, and they fought to preserve local distinctiveness and the right to build their own version of a co-operative and socialist utopia.

In defying the authority of the British empire and the United Kingdom, Irish nationalists drew upon a family of nationalities. Nationality appeared to be simply a claim for autonomy or respect for a community defined at a certain scale. But nationality was also an attempt to define the nature of that community. The hierarchy of distinguishing characteristics could be drawn upon in different ways, having contrasting implications for the future of the polity and the society. Nationality was a discursive prize to be won through essentialist definitions of a particular kind. The struggle, however, was always more than discursive and political, and economic developments blew fair behind some rather than other of these claims at various times. Historians of Ireland such as Hutchinson (1987) have done quite well in capturing public debate as a sort of pendulum between political and cultural nationalism. The question of where authority should lie, and in what sort of future Ireland, is not apparent in such accounts, yet the question of what independence was for was being decided at this time. To illuminate this implicit debate, we have to acknowledge diversity within both cultural and political nationalism. The first had historic, linguistic, and literary streams and the second religious, economic and political. The division was not even this clear and there is, of course, a culture of the economy and a politics of literature, for example. The pendulum theory obscures these questions by making independence appear inevitable, with the cultural swing no more than a compensating response to temporary political failure.

In devising nationalist appeals, then, Irish political and cultural activists were bidding for the right to define the nation. In defending their choices, they referred both above and below the nation. They invoked people's local experiences of solidarity. They also looked to how their actions could be justified before certain international courts of appeal. The rescaling of politics was an attempt to move Ireland from the status of colony and province to that of state. This rescaling was contested and unstable. Thankfully, it still is.

ACKNOWLEDGEMENTS

Thanks for comments and advice to Tristan Clayton, Stuart Corbridge, Millie Glennon, Mike Heffernan, Phil Howell, Denisa Kostovicova, Steve Legg, Denis Linehan, Simon Reid Henry, Ray Ryan, Willie Smyth, Hannah Weston.

REFERENCES

Albright, D. (ed.) 1990: *W. B. Yeats: The Poems*. London: Dent.

Anderson, W. K. 1994: *James Connolly and the Irish Left*. Blackrock: Irish Academic Press.

Æ [G. Russell] 1904a: Physical force in literature. *Dana: An Irish Magazine of Independent Thought*, 1, 129– 33.

Æ [G. Russell] 1904b: Religion and love. *Dana: An Irish Magazine of Independent Thought*, 1, 45–9.

Brennan, T. 1997: *At Home in the World: Cosmopolitanism Now*. Cambridge, MA: Harvard University Press.

Brown, T. 1999: *The Life of W. B. Yeats: A Critical Biography*. Dublin: Gill & Macmillan.

Coffey, D. 1911: Small nationalities. *New Ireland Review*, 34, 277–86.

Coldrey, B. M. 1988: *Faith and Fatherland: The Christian Brothers and the Development of Irish Nationalism, 1838– 1921*. Dublin: Gill & Macmillan.

Connolly, J. 1897a: Socialism and Irish nationalism. *L'Irlande Libre*.

Connolly, J. 1897b: Socialism and nationalism. *Shan Van Vocht*, Jan.

Connolly, J. 1898: Peasant proprietorship and socialism. *Workers' Republic*, 27 Aug.

Connolly, J. 1908: Roman Catholicism and socialism. *The Harp*, Sept.

Connolly, J. 1909a: Sinn Féin, socialism and the nation. *Irish Nation*, 23 Jan.

Connolly, J. 1909b: *Workshop Talks*. Chicago: Charles Kerr & Co.

Connolly, J. 1911: Ireland, Karl Marx and William. *Forward*, 10 June.

Connolly, J. 1914: The solidarity of Labour. *Forward*, 18 Apr.

Connolly, J. 1915: A continental revolution. *Forward*, 15 Aug.

Connolly, J. 1916: The national festival. *Workers' Republic*, 18 Mar.

Connolly, J. 1917[1915]: The reconquest of Ireland. In J. Connolly (ed.), *Labour in Ireland*. Dublin: Maunsel, 217–346.

Cullingford, E. B. 2001: Phoenician genealogies and Oriental geographies: language and race in Joyce and his successors. In E. B. Cullingford (ed.), *Ireland's Others: Gender and Ethnicity in Irish Literature and Popular Culture*. Cork: Cork University Press, 132–60.

De Blaghde, E. 1972: Hyde in conflict. In S. Ó Tuama (ed.), *The Gaelic League Idea*. Cork: Mercier Press, 31–40.

de Vere White, T. 1966[1948]: *Kevin O'Higgins, the Strong Man in the Free State Government 1922–27, his Eventful Life and Mysterious Violent Death*. Dublin: Anvil Books.

Deane, S. 1997: *Strange Country: Modernity and Irish Nationhood since 1790*. Oxford: Clarendon Press.

Dennehy, W. F. 1905: Nationality within the Empire. *New Ireland Review*, 23, 10–16.

Duffy, C. G. 1894[1893]: Books for the Irish people. In C. G. Duffy (ed.), *The Revival of Irish Literature and Other Addresses*. London: T. Fisher Unwin, 35–60.

Duffy, E. 1997: *Saints and Sinners: A History of the Popes*. New Haven: Yale University Press.

Dunleavy J. E., and G. W. Dunleavy (1991) *Douglas Hyde: A Maker of Modern Ireland*. Berkeley, CA: University of California Press.

Eagleton, T. 1998: The Ryan line. In *Crazy John and the Bishop, and Other Essays on Irish Culture*. Cork: Cork University Press, 249–72.

Eglinton, J. [W. K. Magee] 1904: The breaking of the ice. *Dana: An Irish Magazine of Independent Thought*, 1, 11–17.

Eglinton, J. [W. K. Magee] 1906: *Bards and Saints*. Dublin: Maunsel.

Eglinton, J. [W. K. Magee] and Ryan F. [as editors] 1904: Introductory. *Dana: An Irish Magazine of Independent Thought*, 1, 1–4.

Fennell, D. 1985: James Connolly and George Russell. *The Crane Bag*, 9, 56–62.

Fitzpatrick, D. 1978: The geography of Irish nationalism, 1910–1921. *Past and Present*, 78, 113–44.

Foster, R. F. 1991: *W. B. Yeats: A Life*, vol. 1: *The Apprentice Mage*. Oxford: Oxford University Press.

Freeman's Journal 1900: 23 Feb., 6c.

Garvin, T. 1981: *The Evolution of Irish Nationalist Politics*. Dublin: Gill & Macmillan.

Gibbons, L. 1996: Synge, country and western: the myth of the West in Irish and American culture. In *Transformations in Irish Culture*. Cork: Cork University Press, 23–35.

Gill, M. J. 1907: Neo-paganism and the stage. *New Ireland Review*, 27, 179–87.

Goddard, E. 1903: The re-birth of the Irish Celt. *New Ireland Review*, 19, 356–61.

Gregory, A. 1914: *Our Irish Theatre: A Chapter of Autobiography*. London: G. P. Putnam's Sons.

Gregory, A. 1983[1905]: Kincora. In M. Fitzgerald (ed.), *Selected Plays of Lady Gregory*. Gerrards Cross: Colin Smythe, 45–88.

Griffith, A. 1904: *The Resurrection of Hungary: A Parallel for Ireland*. Dublin: James Duffy.

Grote, G. 1994: *Torn between Politics and Culture: The Gaelic League 1893–1993*. Münster: Waxmann.

Hogan, E. M. 1990: *The Irish Missionary Movement: A Historical Survey, 1830–1980.* Dublin: Gill & Macmillan.

Howe, S. 2000: *Ireland and Empire: Colonial Legacies in Irish History and Culture.* Oxford: Oxford University Press.

Howes, M. 1996: *Yeats's Nations: Gender, Class, and Irishness.* Cambridge: Cambridge University Press.

Hutchinson, J. 1987: *The Dynamics of Cultural Nationalism: The Gaelic Revival and the Creation of the Irish Nation State.* London: Allen & Unwin.

Hutchinson, J. H. 1906: Intolerant democracy. *New Ireland Review*, 25, 264–7.

Hyde, D. 1894[1893]: The necessity for de-Anglicising Ireland. In C. G. Duffy (ed.), *The Revival of Irish Literature and Other Addresses*. London: T. Fisher Unwin, 115–61.

Hyde, D. 1967[1899]: *A Literary History of Ireland*. London: Ernest Benn.

Hyde, D. 1986a[1886]: A plea for the Irish language. In B. Ó Conaire (ed.), *Douglas Hyde Language, Lore and Lyrics*. Dublin: Irish Academic Press, 74–80.

Hyde, D. 1986b[1905]: Address [at Carnegie Hall]. In B. Ó Conaire (ed.), *Douglas Hyde Language, Lore and Lyrics*. Dublin: Irish Academic Press, 177–92.

Joy, M. 1905: The Irish literary revival. *New Ireland Review*, 23, 257–66.

Joyce, J. 1959 [1901]: The day of the rabblement. In E. Mason and R. Ellmann (eds), *The Critical Writings of James Joyce*. New York: Viking Press, 68–72.

Kant, I. 1963[1784]: Idea of a universal history from a cosmopolitan standpoint. In L. W. Beck (ed.), *On History*. Indianapolis: Bobbs-Merrill, 11–26.

Kearns, G. 2001a: Time and some citizenship: nationalism and Thomas Davis. *Bullán: An Irish Studies Journal*, 5, 23–54.

Kearns, G. 2001b: 'Educate that holy hatred': place, trauma and identity in the Irish nationalism of John Mitchel. *Political Geography*, 20, 885–911.

Kearns, G. 2002: Ireland after theory. *Bullán: An Irish Studies Journal*, 6, 107–14.

Kearns, G. 2003: Imperial geopolitics. In J. Agnew, K. Mitchell and G. Toal (eds), *A Companion to Political Geography*. Oxford: Blackwell, 173–86.

Kenny, M. 1997: *Goodbye to Catholic Ireland: A Social, Personal and Cultural History from the Fall of Parnell to the Realm of Mary Robinson*. London: Sinclair-Stevenson.

Lane, F. 1997: *The Origins of Modern Irish Socialism, 1881–1896*. Cork: Cork University Press.

Larkin, E. 1979 *The Roman Catholic Church in Ireland and the Fall of Parnell, 1888–1891*. Chapel Hill: University of North Carolina Press.

Levenson, S. 1973: *James Connolly: A Biography*. London: Martin, Brian & O'Keefe.

Lloyd, D. 1993: *Anomalous States: Irish Writing and the Post-colonial Moment*. Dublin: Lilliput.

Lynd, R. W. 1905: The nation and the man of letters. *Dana: An Irish Magazine of Independent Thought*, 1, 371–6.

MacDonagh, T. 1920: *Literature in Ireland: Studies Irish and Anglo Irish*. Dublin: The Talbot Press.

MacNeill, E. 1919: *Phases of Irish history*. Dublin: M. H. Gill & Son.

McKenna, S. 1903: The problem of a world tongue. *New Ireland Review*, 19, 1–10.

Miller, D. W. 1973: *Church, State and Nation in Ireland 1898–1921*. Dublin: Gill & Macmillan.

Moore, G. 1901: Literature and the Irish language. In Lady Gregory (ed.), *Ideals in Ireland*. London: At the Unicorn, 43–51.

Moran, W. P. 1920: Social reconstruction in an Irish state – II. *Irish Theological Quarterly*, 15, 101–12.

Mulligan, A. 1996: The concept of landscape and its use in articulating and narrating an Irish nationalist identity, c.1895–1901. Unpublished M.Phil. thesis, Department of Geography, National University of Ireland (Cork).

Ó Buachalla, S. 1988: *The Letters of P. H. Pearse*. Gerrards Cross: Colin Smythe.

O'Hannay, J. 1906: A word for the Gaelic League. *National Review*, 47, 833–9.

Pakenham, F. 1972 [1935]: *Peace by Ordeal: An Account from First-Hand Sources, of the Negotiation and Signature of the Anglo-Irish Treaty 1921*. London: Sidgwick & Jackson.

Perceval, A. W. 1905: Constructive thought in Ireland. *New Ireland Review*, 23, 360–8.

Ryan, F. 1904: Is the Gaelic League a progressive force? *Dana: An Irish Magazine of Independent Thought*, 1, 216–20.

Ryan, F. 1905: On language and political ideals. *Dana: An Irish Magazine of Independent Thought*, 1, 273–9.

Ryan, F. 1906a: *Criticism and Courage and Other Essays*. Dublin: Maunsel.

Ryan, F. 1906b: Democracy as a discipline. *New Ireland Review*, 25, 331–6.

Seaton, S. 1898: The Gaelic language: its origin and history. *Shan Van Vocht*, 3, 223b.

Sinn Féin 1915: *Sinn Féin in Tabloid Form*. Dublin: Sinn Féin.

Sweetman, J. 1905: Hope within Ireland. *New Ireland Review*, 23, 199–205.

Synan, A. 1905a: The Irish silence. *New Ireland Review*, 22, 267–75.

Synan, A. 1905b: Hatred within the United Kingdom. (Its merits as a practical policy.) *New Ireland Review*, 22, 65–71.

Synge, J. M. 1982a[1902]: The old and new in Ireland. In A. Price (ed.), *J. M. Synge. Collected Works*, vol. 2: *Prose*. Gerrards Cross: Colin Smythe, 383–6.

Synge, J. M. 1982b: Various notes. In A. Price (ed.), *J. M. Synge. Collected Works*, vol. 2: *Prose*. Gerrards Cross: Colin Smythe, 347–51.

Synge, J. M. 1982c: Can we go back into our mother's womb? In A. Price (ed.), *J. M. Synge. Collected Works*, vol. 2: *Prose*. Gerrards Cross: Colin Smythe, 399–400.

Synge, J. M. 1982d[1902]: La vieille littérature irlandaise. In A. Price (ed.), *J. M. Synge. Collected Works*, vol. 2: *Prose*. Gerrards Cross: Colin Smythe, 352–5.

Tymoczko, M. 1994: *The Irish Ulysses*. Berkeley: University of California Press.

van der Veer, P. 1994: *Religious Nationalism: Hindus and Muslims in India*. Berkeley: University of California Press.

Watson, G. J. 1994[1979]: *Irish Identity and the Literary Revival: Synge, Yeats, Joyce and O'Casey*. Washington: Catholic University of America Press.

Webb, A. 1904: The Gaelic League and politics. *Dana: An Irish Magazine of Independent Thought*, 1, 142–4.

Yeats, W. B. 1961a[1900]: The theatre. In *Essays and Introductions*. London: Macmillan, 165–72.

Yeats, W. B. 1961b[1901]: Ireland and the arts. In *Essays and Introductions*. London: Macmillan, 203–10.

Yeats, W. B. 1961c[1910]: J. M. Synge and the Ireland of his time. In *Essays and Introductions*. London: Macmillan, 311–42.

Yeats, W. B. 1964a[1892]: The Irish National Literary Society. In N. Jeffares (ed.), *W. B. Yeats: Selected Criticism*. London: Macmillan, 17–21.

Yeats, W. B. 1964b[1904]: First principles. In N. Jeffares (ed.), *W. B. Yeats: Selected Criticism*. London: Macmillan, 130– 49.

Yeats, W. B. 1999a[1924]: The Irish dramatic movement: a lecture delivered to the Royal Academy of Sweden. In W. H. O'Donnell and D. N. Archibald (eds), *The Collected Works of W. B. Yeats*, vol. 3: *Autobiographies*, New York: Scribner, 410–18.

Yeats, W. B. 1999b[1916]: Reveries over childhood and youth. In W. H. O'Donnell and D. N. Archibald (eds), *The Collected Works of W. B. Yeats*, vol. 3: *Autobiographies*, New York: Scribner, 37–108

Chapter Twelve

British Geographical Representations of Imperialism and Colonial Development in the Early and Mid-Twentieth Century

Robin A. Butlin

Introduction

The aim of this chapter is to review significant links between geographical knowledge and personal and institutional experiences of imperialism, colonialism and development in two critical periods of the twentieth century: *c.*1900–18 and *c.*1935–65. For the first of these periods, the links between geographical knowledge and imperialism in Britain will be explored within contexts of scientific knowledge, increasing modernization of travel and exploration, escalating geopolitical conflict, and changing senses of national identity. Some individuals thought that geography should engage in political commentary and advocacy of commercial development of imperial territories. Other individuals, and at least one institution – the Royal Geographical Society – felt that a scientific society should largely refrain from discussion and publication of such matters.

For the second period, *c.*1935–65, the emphasis will be on the emergence of geography as an increasingly affirming academic discipline with authority and expertise in the field of applied overseas development. Though decolonization was already under way from 1947, surprisingly little commentary and analysis of this process was provided by geographers.

1900–1918

At the beginning of the twentieth century, the publications and records of the Royal Geographical Society (RGS) and other British learned societies reflected the continuing involvement of Britain in the second major phase of European imperialism or 'high imperialism' (*c*.1876–1918). Traditional interests in exploration, discovery and mapping continued, though increasing attention was given to the polar worlds. Racial and ethnological discourses are strongly evident early in the century, and they link in the case of the RGS to a particular kind of learned society: metropolitan, male, 'heroic' militaristic, patriarchal, nationalistic, triumphalist, partially academic, partially scientific, and with links to government and state administration.

Accounts in the *Geographical Journal* for the year 1900, such as H. W. Blundell's 'A journey through Abyssinia to the Nile', with its 'heroic' descriptions of big-game hunting (Blundell 1900), R. Koettlitz's account in racist language of ethnological investigations in Abyssinia (Koettlitz 1900), and H. J. Mackinder's account of 'A journey to the summit of Mount Kenya' (Mackinder 1900) are representative. They are balanced in the same year by a range of scientific, educational and topical material, reflective of the 'new' geography of the time, and its links with modernity. A paper by Vaughan Cornish on sand dunes in the Nile Delta was the opening article for 1900, and material on scientific lake surveys, the Arctic, and the anthropology of British New Guinea (by Alfred Haddon) was published. Short notices of diplomatic and other reports appeared in the 'Monthly Record' section. Some published papers were based on lectures to other societies, notably the British Association for the Advancement of Science, which reached a broader audience and had a more catholic agenda, not least regarding imperial matters. Geographical and other learned societies were affected by, and partly agents of, the complex processes of modernity that were shaping the world (Nash 2000), and engaged with both local and more distant 'spaces' of modernity (Ogborn 1998).

Players, politics, and issues

The opening decades of the twentieth century are characterized by a seeming reluctance at the RGS to acknowledge, debate and publish the imperial discourse in which its officers and Fellows were obviously engaged. An attempt seems to have been made to preserve a 'constitutional' disinterestedness in matters imperial, commercial and political. This seems strange at a time when many senior members were key players in Britain's geostrategic and imperial policies, and when the 'provincial' geographical societies such as Manchester, Liverpool, Hull, Tyneside and Southampton, together with Edinburgh and the branches of the Royal Scottish Geographical Society (founded in 1884), were strong advocates of commercial advancement through empire-based trade (MacKenzie 1995). The interpretation by senior RGS officers of the constitution and charter of the society seemed to act as a means of filtering out any deep analysis of imperial matters, perhaps in an attempt to preserve the image of the RGS as a scientific society, or maybe to preserve the atmosphere of a gentlemen's social club and debating society, where the affairs of the world could be put aside.

The list of RGS presidents from the late nineteenth century to the 1920s reads as a roll-call of the imperial, political, and expeditionary 'great and good': Sir Clements Markham (President 1893–1905), Sir George Goldie (1905–8), Major Leonard Darwin (1908–11), Earl Curzon of Kedleston (1911–14), Douglas Freshfield (1914–17), Colonel Sir Thomas Holdich (1917–19) and Sir Francis Younghusband (1919–22). In addition, though at the beginning of the twentieth century he was a relatively junior member, a major and influential figure was H. J. Mackinder. Markham had explored in Peru, specifically to find cinchona trees (whose bark had anti-malarial properties) for transfer for commercial production in India; he had reorganized the geographical section at the India Office, and promoted Scott's British National Antarctic expedition of 1899–1902 and the fatal expedition of 1910–12. Goldie had been a major player in Britain's trading, settlement and geopolitical activities in West Africa, especially in what became Nigeria, and attended the Berlin West Africa Conference of 1884–5 as an unofficial adviser on 'Niger matters' to the British delegation (Flint 1960: 68). Darwin was an army engineer and administrator who had taken astronomical observations in the Transit of Venus observation expeditions of 1874 and 1882 (Mill 1930: 139). Curzon had been a Conservative MP from 1886, Viceroy of India (1898–1905), Lord Privy Seal (1915–16) and Foreign Secretary (1919–24). All sought an expansion of geographical education in the Civil Service, armed forces, public schools and Oxford and Cambridge universities, especially after Inspector of Geographical Education J. S. Keltie's report to the society in 1886 on improvement in geographical education.

Curzon's views on geography and empire are evidenced in a speech at the RGS annual dinner in 1912, whose principal guest was Prime Minister H. H. Asquith. Curzon saw Asquith's agreement to attend the dinner as:

> a recognition by the first public man in England of the part geographers have played in consolidating and organizing the British Empire. Without the geographers that Empire never would have been created...Everywhere throughout the world you still find them exploring and surveying unknown lands, demarcating boundaries, planting the seeds of commerce, pacifying savage tribes, and extending the frontiers of civilization. (Curzon 1912: 103)

Curzon solicited a supportive response from Asquith, who did not, however, rise to the bait:

> I will not, however strong the temptation may be, enter on the interesting and much disputed controversy as to the relation between politics and Geography...I think I may truly say that if the modern politician requires, as undoubtedly he does require if he is worthy of his task, to be equipped with geographical knowledge, on the other hand you geographers must admit that geography owes much to the modern politician. (Asquith 1912: 106)

Of the remainder of those RGS presidents listed above, Freshfield was a distinguished mountaineer and active promoter of geographical education, and Holdich was a military surveyor, expert on the survey and arbitration of international boundaries, who had been attached to the Survey of India and involved in intelligence-gathering activities on the Indian frontier in the late nineteenth century. His general book on India in 1904 seems, notwithstanding its overly strong emphasis on Afghanistan, to have been the main English-language geography text until Spate's 1954 *India and Pakistan*. Younghusband, a protégé of Curzon, had been a major player in geographical exploration and intelligence in India and Central Asia in the late nineteenth and early twentieth centuries, though as President he showed much interest in representations of landscape, which according to Mill 'afforded a welcome relief to the useful but heavy reports on new air-routes, and on the new methods of surveying' (1930: 209).

These and other RGS officers and members symbolized geographical knowledges of military activity and strategy, exploration, survey and mapping, ethnographic and environmental studies (notably of acclimatization, with strong racial overtones), and more general accounts of far-off places. Driver has argued, in relation to the exploration role of the society, that 'the culture of exploration was heterogeneous, and the knowledge it produced took a great variety of forms. In this context, what constituted legitimate knowledge was always a matter of contention' (2001: 28–9). Even so, the small number of published papers on geopolitical matters, including, for example, the Anglo-Boer War (a turning point in imperial experience) is surprising. Brief mention is made in Goldie's comments on the inadequacy of British map resources for the conduct of the war (Goldie 1907). It could be argued that by the beginning of the twentieth century the major imperial and geopolitical issues in Asia and Africa were largely over. But Curzon had not abandoned his concern about Russia menacing India, as he showed at the Anglo-Russian Convention of 1907 (Gilmour 1994: 377), and there was still much geographical interest in empire, increasing during the First World War. The RGS continued to be directly and indirectly involved in arbitration in sensitive international boundary disputes, and its military Fellows engaged in action in Europe, Africa and elsewhere. In *The Indian Borderland 1880–1900* Holdich (1900: 366) indicated the ends that could be served by the acquisition of geographical information through exploration and trigonometrical survey, and maintaining a military presence, not least in Afghanistan, to offset what he still saw as a potential Russian threat to India.

A leading figure in the RGS and in the development of geopolitical thought at this time was H. J. Mackinder. His account of ascending Mount Kenya in 1900 has been analysed by Kearns in comparison with Mary Kingsley's accounts of her travels in West Africa (Kearns 1997). Mackinder is a significant figure in the analysis of the links between geography, imperialism and modernity at that time. Ó Tuathail, like others, ascribes to Mackinder (elected FRGS in 1885) the role of an active and effective promoter of a new geography linked to a renewed imperialism, suggesting that 'the metaphor of geography as a bridge over various abysses that were said to characterize British culture (and modernity in general) was a popular one within the RGS', giving the discipline 'an exceptionalist myth' and the ability 'to arrest the relative decline of British power and renew the idea of empire' (Ó Tuathail 1996: 86). This might be achieved through advancement of geographical education in the civil and armed services, schools and universities, and the extensive use of maps, visual aids (including lantern slides) and exhibitions (Ryan 1997). Kearns has also analysed Mackinder's role as imperial propagandist and activist at a time of imperial 'crisis' from *c*.1897 to 1914, his closed space theory and biological metaphors contributing to 'the paranoid style in *fin de siècle* British politics' (1993: 29). Mackinder argued that an imperial protectionist tariff should be created, and the empire should be sustained in order to maintain Britain's military and naval strength and its trade links (Kearns 1993: 22). Mackinder had the opportunity to put his geographical ideals into practice when he was appointed, with the help of Curzon as Foreign Secretary, as high commissioner to southern Russia in 1919 (Livingstone 1992: 195).

The renewal of empire was thus an important theme in the writings of RGS members in the early twentieth century, and in discourses of other learned societies published in the *Geographical Journal*. In his Presidential Address to the Bradford Meeting of the British Association in 1900, on 'Political geography and the British empire', Sir George S. Robertson took a less cautious line on imperialism than is usually found in *Geographical Journal* papers; the British Association itself had a science-related ideological and imperial mission (Dubow 2000: 69):

> Pure geography, with its placid aloofness and its far-stretching outlook, combined sometimes with a too rigid devotion to the facts and conclusions of strict geographical research...there is nothing foreign to geographical thought in the association of geography and patriotism...That is my justification and my apology for taking political geography and the British Empire

as my subject, if justification and apology seem to anyone to be necessary. (Robertson 1900: 447)

The need for the renewal of imperial enthusiasm was tied to a mood of pessimism about the empire, related by Hyam to the effect of the Anglo-Boer War, with its failings of military strategy, inadequate maps, and deficiencies in the health of British soldiers involved, and to the broader questioning of Britain's ability to control large and far-flung parts of the world: 'A drive for "national efficiency" was given an irresistible momentum within the context of the need to re-engine the empire' (Hyam 1999a: 50). This mood, accompanied and partly fed by such radical critiques of empire as J. A. Hobson's 1902 *Imperialism*, led ultimately to a more moderated, constructive and humanistic view of the empire in the future, itself facilitated by the increased Liberal power, against that of the Unionists, after the election of January 1906. Political changes provided an opportunity for militant imperialists such as Curzon to revive their imperial dreams, at least for a time, and especially during the First World War: 'The "new imperialists" of the 1890s (who were now older imperialists)...were spirited into the government, which became less Liberal, and more congenial company for the imperialists, as the war went on' (Porter 1996: 243). Curzon was one such member of Lloyd George's government.

This temporary swing of the pendulum towards imperial renewal is picked up in a slight change of attitude to political discussion in RGS meetings. The conventional conservative view is shown in Sir George Goldie's speech on 11 February 1901 to the RGS meeting in commemoration of the late Queen Victoria. Speaking of Africa, he said that:

> our political expansion there is the recent outcome of a continuous and extremely heated international scramble...It seems to me that one cannot now deal in any adequate way with that controversial history before a Society which, though never forgetting that it is a British Society, invites geographers of all nationalities to its meetings, and pursues the even tenor of its scientific and therefore cosmopolitan way in the acquisition, encouragement, and diffusion of geographical knowledge. (Goldie 1901: 236–7)

A slightly more liberal view is taken by Holdich in his presidential address in 1919: 'We have recognized this year the necessity for some relaxation of the old but unwritten law which governed our meetings before the war, that political allusions were to be discouraged' (Holdich 1919: 6). The reason for this unwritten law is not entirely clear. It might relate to perceived conflict with the chartered ideals of the society, or be a response to the expectations of their political masters and the War Office that no strategic or politically sensitive information should be brought before the society. Compared with France, however, there is surprisingly little critique of empire by British geographers in the early part of the twentieth century. Sir Francis Younghusband affords a brief glimpse in a speech at the RGS anniversary meeting in 1919. Speaking of the lack of enthusiasm for geography in schools, he quotes some 'remarks' by 'Mr Mackinder':

> In this country we value the moral side of education, and it is perhaps intuitively that we have neglected materialistic geography. Before the war not a few teachers, within my knowledge, objected to geography on the ground that it tended to promote Imperialism, just as they objected to physical drill because it tended to militarism. (Younghusband 1919: 7)

Some sense of geographical thought on the eve of the First World War can be obtained from Rudyard Kipling's lecture to the RGS on 17 February 1914 on 'Some aspects of travel'. Kipling reviews the modernization of travel in a world already changing but braced for further change, covering the psychology and pressure of travel, gifts of visualization on expeditions, the use of atlas maps and the notion of mental maps and pictures of travel, the traveller and nostalgic smells (geographies of the senses), and, in conclusion, prospects for the future:

> Month by month the Earth shrinks actually, and what is more important, in imagination.... Only the spirit of man carries on, unaltered and unappeasable. There will arise – they are shaping themselves even now – risks to be met as cruel as any that Hudson or Scott faced; dreams as world-wide as Columbus or Cecil Rhodes dreamed, to be made good or to die for; and decisions to be taken as splendidly terrible as that which Drake clinched by Magellan, or Oates a little further south. There is no break in the line, no loads are missing; the men of the present have begun the discovery of the new world with the same devoutly careless passion as their predecessors completed the discovery of the old. (Kipling 1914: 374–5)

The imperial undertones are clear from Kipling's references to Rhodes and in comments in discussion by other members of the society. The meeting was chaired by Curzon, and the vote of thanks proposed by Lord Bryce, former explorer and subsequently British ambassador to the USA. The lecture is, not surprisingly, heavily masculinist in tone, as was most RGS discourse at this time. Curzon himself had originally opposed the election of women as Fellows, and then changed his mind in 1913, while at the same time opposing women's suffrage, arguing 'not very persuasively that to give women "a share in the Sovereignty of the country and the Empire" was a wholly different thing from giving them a voice in a society which existed for "nothing more formidable or contentious than the advancement of a particular department of human knowledge"' (Gilmour 1994: 401).

Lycett suggests that in his lecture Kipling was implying that 'Writers and explorers were in the same business of charting the new world . . . Only two months earlier, Apsley Cherry-Garrard had sent him a copy of *Kim* that had been read by members of Captain Robert Scott's recent ill-fated expedition to the South Pole. In June Sir Ernest Shackleton visited Bateman's [Kipling's house] to discuss his own Antarctic voyage' (Lycett 1999: 438). The society's agenda at that time was of course wider than Kipling's, and included the promotion of geography through education in schools and universities and the evaluation of areas for future settlement (including the question of acclimatization), at a time of increasing population and emigration from industrializing Europe. The RGS's imperial understanding accepted a clear role in empire through training in the compilation and use of maps and of up-to-date geographical information. A. J. Herbertson's address to the British Association at Sheffield in 1910 on 'Geography and some of its present needs' put forward 'The need of the geographical factor in Imperial Problems', arguing for a geographical-statistical department of government that would supply relevant data to politicians (Herbertson 1910: 477–8).

Geography as science

The definition and promotion of geography as a science was a major issue for the RGS, witnessed in the debates of the 1890s between the traditionalist map-makers, navigators and explorers, and Mackinder, Freshfield and Galton, advocates of the advancement of geography as a science (Livingstone 1992: 193). The increasing interest in geography as an environmental as well as a cartographic science offered on the one hand opportunities for scientific theory to continue to underpin imperialism and overseas settlement, and on the other a tendency to reduce if not to block out the stronger urges of geographers for engagement with political policy. Smith (1994, 1999) has shown these tendencies at work in the United States in the ideology of the leading geographer Isaiah Bowman, arguing that American geographers were:

> squeezed between two contradictory forces. On the one hand geography was steeped in the narrow patriotism that tragically accommodated and only belatedly displaced a politically expedient if utterly misguided environmental determinism . . . On the other hand, however, the emerging discipline of geography already conceived itself widely as scientific, eschewing anything that could be conceived as a political stance. The turmoil provoked by this contradiction between politics and science etches much of the identity of American geography in the first half of the twentieth century, not least

Bowman's efforts to write a political geography that was, by his judgement, non-political. (Smith 1999: 13–14)

Smith argues that geography was isolated from the modernist 'ferment' of the early twentieth century, with a consequential delay in engagement with the wider world until the 1960s.

Could the same be said of British geography, or at least of English metropolitan geography? There were, of course, differences in political ideology, and British geography did not spring from the same powerful roots in physical geography as did its American equivalent. It could also be argued that Scotland, through its own geographical society, and the 'provincial' English societies for a shorter time, embraced wider global and political perspectives than the RGS (MacKenzie 1995; Bell 1995; Maddrell 1997). Women had been members of the Scottish Geographical Society since its inception, unlike the RGS, which had not admitted women as Fellows until 1913 (Bell and McEwan 1996). The Scottish and English provincial geographical societies also vigorously supported commercial geography, largely ignored by the RGS (Lochhead 1984; MacKenzie 1995; Barnes 2002). There was also a difference in membership composition of the societies, the RGS having suffered a reduction in the number of scientists (as evidenced by overlapping membership with the Royal Society) and a persistence of aristocratic and titled gentry and naval and military officers at the turn of the century, whereas the Scottish society had more eminent scientists (Lochhead 1984). The contributions to meetings and publications of the RGS and to Section E of the British Association for 1906–11 presented by Howarth (1951: 153) show similar contrasts (see table 12.1).

Table 12.1.   Contributions to meetings and publications of the RGS and to Section E of the British Association for 1906–1911

	RGS (%)	British Association (Section E) (%)
Explorations	57	22
Physical geography	29	43
Economic and social geography	7	23
Mathematical and cartographical geography	3	12
Anthropological and biological	4	(in other BA sections)

Source: Howarth 1951: 153.

There was, however, a clear view from RGS officers such as H. R. Mill that geography's scientific aspects were increasingly important, and that the scientific quality of papers read to and published by the society was improving. Of 1905–14 Mill states: 'The papers read to the Society each year were no more numerous than before, but the quality was higher and the intensive study of comparatively small areas allowed the great principles of science to outshine the daily details of journeys in strange places, among unfamiliar plants and animals and incomprehensible people' (Mill 1930: 187). Major scientific issues debated included questions of climate, race, and potential areas for colonial settlement, technical questions of mapping by land and air survey, and aspects of polar exploration (Livingstone 1992, 2002; Cameron 1980). In terms of the society and its local space of modernity, Mill points out that its newly acquired house at Kensington Gore had become increasingly accessible through the 'introduction of tube railways and the universal use of motor buses and taxis' (Mill 1930: 182), to which might be added the earlier use by the society of the telephone and typewriter, and, of course, of lantern slides. In 1902 Holdich had extolled, in an address to the British Association, the importance of the telegraph in the new scientific geographical order (1902: 518).

The RGS's activities and publications for 1914–18 show a small amount of published work on the geographical consequences of war. No mention is made of the Russian Revolution, and Hilaire Belloc's paper on 'The geography of the war' (1915) was mainly on topographical obstacles in northeastern France during the early stages of the conflict. The involvement of the RGS's resources and personnel with wartime activity, particularly through the Geographical Section of the General Staff (GSGS) of the War Office and through war cartography (Heffernan 1996: 507), differed from that of other countries, notably France and the United States (conspicuously Isaiah Bowman), where 'leading geographical experts' were used in 'a much more substantive political capacity' (Heffernan 1996: 520).

British Geographies of Development c.1935–1965

Historians of British imperialism indicate that after the First World War British imperial policy became less aggressive and more concerned with development. This had much to do with the costs of empire, seen against the large debt resulting from the war. Alongside continuing concerns in Ireland, India and the Middle East, policies were developed for more enlightened administration, including the Dual Mandate, and early concepts of self-government and Commonwealth development. Owen (1999: 193) has argued that such features were pragmatic responses to financial and geopolitical crises rather than a major re-framing of ideologies of empire. Such policies, as they progressed in the 1930s, with an emphasis on Africa, peasant agricultural development, and the lessening of the white settler impact in colonies, led, *inter alia*, to a renewed scientific interest in land use, agriculture, population distribution, urbanization, health, and migration. Politically, what has been described as a 'rough consensus' (Marshall 1996: 84) on imperial development developed between Labour and Conservatives, now the main political parties.

Legislation, including the 1929 Colonial Development Act (which provided £1,000,000 for economic development and trade with Britain) and the 1940 Colonial Development and Welfare Act, pointed the way. Two major surveys, published in 1938, were commissioned as a basis of future African development; *African Survey*, produced under Lord Hailey, formerly Governor of the Punjab and the United Provinces in India, and E. B. Worthington's *Science in Africa: a review of scientific research relating to Tropical and Southern Africa*. By this time, many universities and university colleges in Britain had established geography departments, albeit with small numbers of staff and students. The time seemed ripe for greater involvement by geographers in the affairs and science of a shrinking world, including questions of empire, but while much important geographical research was undertaken in this period, surprisingly little was written on decolonization. A paradox here is that post-war British strategy had been influenced by the work of Mackinder. Hyam highlights links between Mackinder's work on 'The geographical pivot of history' (1904) and *Democratic Ideals and Reality* (1919) and British geopolitical and decolonization policy, through the agency of L. S. Amery, and Ernest Bevin's overt reference to Mackinder's concepts of a world-island being employed in the Cold War by the Politburo (Hyam 1999b). Mackinder led in theory, but where were the geographers to follow his ideas?

In practice, the experiences of a number of geographers in various parts of the Tropics during the Second World War did equip them with the curiosity, expertise and regional experience to advance the modernization of various parts of the world, including imperial territories (Farmer 1983). The RGS 'United Kingdom Geographers in the Second World War' project exemplifies the roles undertaken by UK geographers in wartime: 'political geographers contributed in Dominion, Colonial and Foreign Office activities and economic geographers in economic warfare' (Balchin 1987: 160). E. H. G. Dobby is known to have played a counter-propaganda/intelligence role in Malaya-Indonesia. Others named in the archive or otherwise known to be active in India, Africa and the Middle East, included Charles Fisher, W. B. Fisher, B. L. C. Johnson, O. H. K. Spate, J. T. Coppock, A. T. A. Learmonth, and R. O. Buchanan. Geographers also researched and published on colonial territories in the Geographical Handbooks produced for the Naval Intelligence Division of the Admiralty during the

Second World War (Balchin 1987). These were produced from centres at Oxford and Cambridge, and included coverage of French West Africa, French Equatorial Africa, the Belgian Congo, Morocco, Algeria, Tunisia, Palestine–Transjordan, Syria, Western Arabia, Iraq and Persia (all these from the Oxford centre, headed by K. Mason) and of Indo-China, the Netherlands East Indies, China, and The Pacific Islands (from the Cambridge centre, headed by H. C. Darby). Steel pointed out that the Handbooks for French West Africa, French Equatorial Africa and the Belgian Congo 'drew heavily on the work of geographers, though rarely upon their fieldwork, for which in any event there was little opportunity during the war years' (Steel 1964: 4).

Most of the significant work in the post-war period on the developing world came from this generation of geographers, and subsequently from their students, with their researches increasingly published in the *Geographical Journal*. There was initially little involvement with geopolitical questions, with notable exceptions. O. H. K. Spate, for example, had worked for the Muslim case in the determination of the India/Pakistan boundary in the period before independence in 1947: 'I was not above the battle but in the thick of it, and being a political animal I thoroughly enjoyed it' (Spate 1947: 201–2). Spate's *India and Pakistan: a general and regional geography* (1954), extended by Spate and Learmonth in 1967 as a third edition, was one of a number of important post-war regional geographies. Charles Fisher also contributed to the post-war study of South-East Asian geopolitics. Fisher's 'The Vietnamese problem in its geographical context' includes analysis of the ideologies behind nationalist movements in French Indo-China and the location of these processes within a broader Asian geopolitical perspective (Fisher 1964, 1965).

Steel has suggested that in the 1930s the 'geographers of the time recognized the need to stress the relevance of geography in both national affairs and international relations' (Steel 1987: 66–7). Freeman, however, alleges that few papers in human geography were published in the *Geographical Journal* at that time (1980: 19), though the American *Geographical Review* did publish such material. The earliest publications of the new Institute of British Geographers, founded in 1933 as an academic breakaway group from the RGS, were monographs: individual papers in the IBG's *Transactions* came after the Second World War, and thematic collections much later. Two explanatory factors cited by Forbes (1984) for the limited amount of colonial geographical work were the small numbers of academic geographers and the tightness of funds and opportunities, so that only a few books, such as Stamp's 1929 *Asia*, were produced.

The period before and following the Second World War is a significant yet neglected period in the historiography of geography, coming between the imperial geographies of the earlier part of the century and the rapid development of geography as a spatial science in the 1960s and 70s. There was however post-war advocacy for stronger colonial engagement with colonial development. In *Transactions* in 1948 R. J. Harrison Church made a case for a more vigorous colonial geography, influenced by his academic experience in France (Harrison Church 1948: 15), and substantiated in his *Modern Colonization* (1951). Academic appointments were few and institutes of colonial geography did not exist in Britain, as they did in France, the Netherlands, Belgium and Portugal. Harrison Church considered various aspects of the geography of colonization and political geography, including boundaries, but does not really attempt the kind of political analysis of metropolitan and colonial aspirations for independence that one might have anticipated.

The initiatives in this period were therefore in researching development issues, but the significance and the rapidity of decolonization between 1947 and 1960, notably in Africa (fig. 12.1), is rarely reflected in geographical

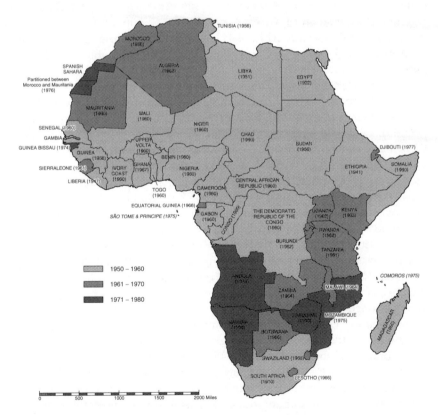

Figure 12.1.   Africa: Chronology of independence
Source: Chamberlain 1999.

publications, a feature that continued into the 1980s (Potter and Unwin 1988).

Geographers in Africa and Asia

As we have seen, there were professional geographers working on Africa and Asia from just before the Second World War, and thus on parts of the empire about to achieve independence. These contributions were significant. A major piece of research was the interdisciplinary Ashanti Survey part-funded by the Colonial Office (Fortes, Steel and Ady 1948). Major contributions were made in the development geography of Africa and India by R. M. Prothero, R. W. Steel and Learmonth, and others from the Liverpool University Geography Department, on applied geography, including health (malaria eradication) and environmental issues. Other contributions included those of G. J. Butland on Latin America, S. J. K. Baker on Uganda, Buchanan and J. C. Pugh on Nigeria, Stamp on Asia and Africa, B. H. Farmer on Ceylon/Sri Lanka, Harrison Church and A.T. Grove on West Africa, and Dobby on South-East Asia. A bibliographical review was published by Steel (1964).

Steel regretted the 'insignificant attention given to the political geography of tropical Africa... This is unfortunate at a time of great political change and experiment in areas like East Africa, the Zambezi basin and Nigeria, where there should have been the same careful analysis by geographers as there has been by historians, economists and political scientists' (Steel 1964: 15). He acknowledged, however, that in the 1950s and 1960s, there were geographers writing on political matters, including Buchanan (1953) on South Africa and Nigeria, Michael Barbour (1959) on international boundaries in Africa and on the Sudan, and Victor Prescott (1959) on boundaries in Nigeria; many of these were published in the *Geographical Review* (also Hamdan 1963). Funding for research was a problem, but progress was made with the establishment of the Colonial Social Science Research Council in 1944, of which geographer F. Debenham became a member in 1948. The CSSRC Committee on Anthropology and Sociology commissioned regional surveys, and a small number of geographers were given support. Gilbert and Steel commented on the general neglect of geography by the CSSRC, on the grounds that it appeared not to be a 'main branch of social science' (1945: 120). The CSSRC seems to have worked on an *ad hoc* basis, with social anthropology as the predominant discipline 'because its representatives had both the necessary field-work experience before the war and the skill to establish the subject in university circles' (Lee 1967: 198–9). A list of social scientists in British Africa in 1956 compiled by Lee from CSSRC data (figure 12.2) shows the numerical predominance of social anthropology and the small number of representatives of geography (Lee 1967: 24).

Anthropologists, whose subject and institutional history are close to those of geography, seem to have been much more in the government's eye up to and beyond the Second World War in terms of imperial and colonial research (Kuklick 1991). Neither Hailey's *African Survey* – which recommended research funding for development studies under the aegis of the Colonial Development and Welfare Act 1940 – nor Worthington's *Science in Africa* had taken much note of geographical research. Their production reflected a rethinking of imperial aims in the 1930s, partly

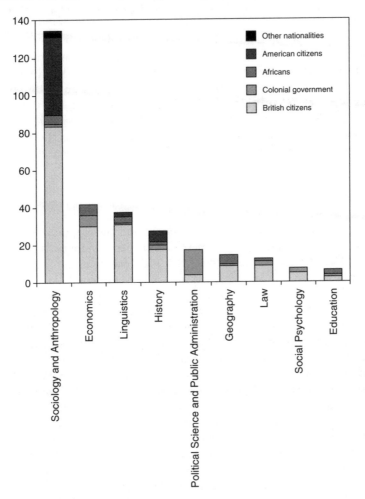

Figure 12.2. Social scientists in British Africa, 1956
Source: J. M. Lee 1967.

associated with Rhodes House, founded in 1929 in Oxford, towards Africa and away from the increasingly American-influenced Far East and Pacific (Rich 1986: 146). *African Survey* was overseen by the Royal Institute of International Affairs and funded by Rhodes Trustees and the Carnegie Corporation in New York. It examined not only past and projected British policies in Africa south of the Sahara, but also those of France, Belgium and Portugal, and addressed questions of race, law, education and technical topics including soil erosion (Louis 1999: 30–2). Little attention was paid to geographical literature (Gilbert and Steel 1945: 121) and there were only three maps in the text, two of them at small scale from W. Fitzgerald's 1934 textbook *Africa*, a remarkably limited provision for a book of over 2,000 pages.

There are clear international contrasts in writing about colonialism and decolonization. Colonialism in France, for example was 'more deep-rooted and violent than the colonialism even of the British', and 'the French were correspondingly more reluctant than the British to let go of Empire after World War Two' (Corbridge 1993: 178). This might explain the stronger interest by French geographers in colonialism and decolonization, and also the supportive role in Portuguese geopolitical writing for the tenacity of Portuguese imperialism in Angola and Mozambique (Power 2001).

Conclusion

During the twentieth century geographers and geographical societies in Britain had a fluctuating involvement with issues of imperialism, colonialism, and ultimately (though belatedly) with decolonization. It has been argued here that the imperial mission was maintained up to the First World War, when there was a short acceleration of political interest, and further involvement by geographer-politicians such as Curzon and Mackinder. Generally, however, geographical literature, especially the *Geographical Journal*, underplayed the reporting and critique of the imperial/colonial mission. In the period immediately before and after the Second World War, important work was carried out by geographers, addressing questions of agriculture, population, urbanization, health, and transport, within a framework of development geography. For Watts, however, 1945–60 (phase 3 of his 'map of development discourse') was a time when 'geographers had little or nothing to say' on the 'deeper and darker political ailments' of brutal totalitarian regimes in Africa (Watts 1993: 179). This thesis is probably less applicable to work on Latin America and India, both of which seem to have received published analyses of greater theoretical import. In the case of decolonization it may well be that speed was an important factor, so that the end of empire politically was far ahead of its intellectual analysis. Lack of early attention to decolonization may have been a peculiarly British phenomenon. French geographers seem to have been much more directly concerned with the French imperial and colonial process from start to finish, evidenced for the last quarter of the twentieth century, for example, in the pages of the journal *Hérodote*, founded in 1976 (Heffernan, 1994, 1995, 1996; Claval 2000; Hepple 2000).

From the late 1960s (especially after 1968) and the early 1970s, new influential schools of critical human geography began to emerge, reflected in journals such as *Antipode* from the late 1960s and *Political Geography Quarterly* (later renamed *Political Geography)* in the early 1980s. Extensive critiques of Western ideologies and theories of modernization, characteristic of post-war models and policies relating to Africa, Asia and Latin America, were undertaken, viewed against a geopolitical background of the growth of the power of the USA and the USSR, the Cold War, and increasing 'political turbulence in the regions of the Third World, with decolonization struggles, liberation wars and revolutionary upheavals' (Slater 1997: 643). 'Post-developmentalism' was characteristic of the mid-to late 1990s, and offered radical critiques of developmentalism (Corbridge 1999; Watts 1993).

Connections with imperial pasts in many parts of the world, including Afghanistan, and India and Pakistan (in relation to the Kashmir question, American-dominated geopolitics, and internal questions of national identity) continue doggedly and dramatically into the twenty-first century. The historical and present issues tied to imperialism are obviously complex, and need deconstructing with care, perhaps with greater aid in future from perspectives of individual biographical and national identity. The unanswered questions about pre- and post-colonial deconstructions of the global knowledges and writings of imperialism are complex and difficult (Sidaway 2000), and await our further attention.

ACKNOWLEDGEMENTS

I am deeply indebted to Professor R. Mansell Prothero and Professor Andrew Learmonth for their help in reconstructing aspects of British geographical connections with empire in the period from *c*.1945, and for their comments on an early draft of this chapter. Dr David Preston and Dr Marcus Power of the School of Geography, University of Leeds, have also made helpful comments, as have the editors of this book: I am grateful to all of them. I acknowledge the help and support of the Director, Keeper, Archivist and Librarian of the Royal Geographical Society for facilitating access to research material, and of the British Academy and the Royal Geographical Society through research grants that supported some of the research included in the chapter.

REFERENCES

Asquith, H. H. 1912: Speech at the Annual Dinner of the RGS. *Geographical Journal*, 40, 105–8.

Balchin, W. G. V. 1987: United Kingdom geographers in the Second World War: a report. *Geographical Journal*, 153(2), 159–80.

Barbour, K. M. 1959: Irrigation in the Sudan: its growth, distribution and potential extension. *Transactions Institute of British Geographers*, 26, 243–63.

Barnes, T. J. 2002: Performing economic geography: two men, two books, and a cast of thousands. *Environment and Planning A: Environment and Planning*, 34, 487–512.

Bell, M. 1995: Edinburgh and empire: geographical science and citizenship for a 'new' age. *Scottish Geographical Magazine*, 111(3), 139–49.

Bell, M., and McEwan, C. 1996: The admission of women Fellows to the Royal Geographical Society, 1882–1914: the controversy and the outcome. *Geographical Journal*, 162(1), 295–312.

Belloc, H. 1915: The geography of the war. *Geographical Journal*, 45(1), 1–15.

Blundell, H. W. 1900: A journey through Abyssinia to the Nile. *Geographical Journal*, 15(2), 97–120.

Buchanan, K. M. 1953: The northern region of Nigeria: the geographical background of its political duality. *Geographical Review*, 43, 451–73.

Cameron, I. 1980: *To the Farthest Ends of the Earth*. London: Royal Geographical Society/Macdonald.

Chamberlain, M. E. 1999: *Decolonization*. Oxford: Blackwell.

Claval, P. 2000: *Hérodote* and the French Left. In K. Dodds and D. Atkinson (eds), *Geopolitical Traditions: A Century of Geopolitical Thought*. London: Routledge, 239–67.

Corbridge, S. 1993: Colonialism, post-colonialism and the political geography of the Third World. In P.Taylor (ed.), *Political Geography of the Twentieth Century: A Global Analysis*. London and New York: John Wiley, 173–205.

Corbridge, S. 1999: Development, post-development, and the global political economy. In P. Cloke, P. Crang and M. Goodwin (eds), *Introducing Human Geographies*. London: Arnold, 67–75.

Curzon, G. N. 1912: Speech at the Annual Dinner of the RGS, *Geographical Journal*, 40, 101–5.

Driver, F. 2001: *Geography Militant: Cultures of Exploration and Empire*. Oxford: Blackwell.

Dubow, S. 2000: A commonwealth of science: the British Association in South Africa, 1905 and 1929. In S. Dubow (ed.), *Science and Society in Southern Africa*. Manchester: Manchester University Press, 66–99.

Farmer, B. H. 1983: British geographers overseas. *Transactions Institute of British Geographers*, NS 8(1) 70–9.

Fisher, C. A. 1964: *South-East Asia: A Social, Economic and Political Geography*. London: Methuen.

Fisher, C. A. 1965: The Vietnamese problem in its geographical context. *Geographical Journal*, 131, 502–15.

Flint, J. E. 1960: *Sir George Goldie and the Making of Nigeria*. Oxford: Oxford University Press.

Forbes, D. K. 1984: *Geography of Underdevelopment*. London: Croom Helm.

Fortes, M., R. W. Steel, and P. Ady 1947: Ashanti survey 1945– 46: an experiment in social research. *Geographical Journal*, 110, 149–77.

Freeman, T. W. 1980: The Royal Geographical Society and the development of geography. In E. H. Brown (ed.), *Geography Yesterday and Tomorrow*. London: Oxford University Press, 1–99.

Gilbert, E. W., and R. W. Steel 1945: Social Geography and its place in Colonial Studies. *Geographical Journal*, 106, 118–31.

Gilmour, D. 1994: *Curzon*. London: Macmillan.

Goldie, G. T. 1901: Progress of exploration and the spread and consolidation of the empire in America, Australia, and Africa. *Geographical Journal*, 16(3), 231–40.

Goldie, G. T. 1907: Geographical ideals. *Geographical Journal*, 19(1), 1–14.

Hailey, Lord 1938: *An African Survey: A Study of Problems Arising in Africa South of the Sahara*. London: Oxford University Press.

Hamdan, G. 1963: The political map of the new Africa. *Geographical Review*, 50, 21–40.

Harrison Church, R. J. 1948: The case for colonial geography. *Transactions Institute of British Geographers*, 14, 15–25.

Harrison Church, R. J. 1951: *Modern Colonization*. London: Hutchinson.

Heffernan, M. 1994: The science of empire: the French geographical movement and the forms of French imperialism. In A. Godlewska and N. Smith (eds), *Geography and Empire*. London: Blackwell, 92–114.

Heffernan, M. 1995: The spoils of war: the Société de Géographie de Paris and the French empire, 1914–1919. In M. Bell, R. A. Butlin and M. J. Heffernan (eds), *Geography and Imperialism 1820–1940*, Manchester: Manchester University Press, 221–64.

Heffernan, M. 1996: Geography, cartography and military intelligence: the Royal Geographical Society and the First World War. *Transactions Institute of British Geographers*, 21(3), 504–33.

Hepple, L. 2000: *Géopolitiques de gauche:* Yves Lacoste, *Hérodote*, and French radical geopolitics. In K. Dodds and D. Atkinson (eds), *Geopolitical Traditions. A Century of Geopolitical Thought*. London: Routledge, 268–301.

Herbertson, A. J. 1910: Geography and some of its present needs. *Geographical Journal*, 36, 468–79.

Holdich, T. H. 1900: *The Indian Borderland 1880–1900*. London: Methuen.

Holdich, T. H. 1902: The progress of geographical knowledge. *Scottish Geographical Magazine*, 18, 505–25.

Holdich, T. H. 1919: Address at the Anniversary Meeting, 2 June 1919, Colonel Sir Thomas H. Holdich, President. *Geographical Journal*, 54(1), 1–12.

Howarth, O. J. R. 1951: The centenary of Section E (Geography) in the British Association. *Scottish Geographical Magazine*, 67(3–4), 145–80.

Hyam, R. 1999a: The British empire in the Edwardian era. In J. M. Brown and Wm. R. Louis (eds), *The Oxford History of the British Empire*, vol. 4: *The Twentieth Century*. Oxford: Oxford University Press, 47–63.

Hyam, R. 1999b: The primacy of geopolitics: the dynamics of British imperial policy, 1763–1963. *Journal of Imperial and Commonwealth History*, 27(2), 27–52.

Kearns, G. 1993: *Fin-de-siècle* geopolitics: Mackinder, Hobson, and theories of global closure. In P. J. Taylor (ed.), *Political Geography of the Twentieth Century*. London: Bellhaven Press, 9–30.

Kearns, G. 1997: The imperial subject: geography and travel in the work of Mary Kingsley and Halford Mackinder. *Transactions Institute of British Geographers*, NS 22(4) 450–72.

Kipling, R. 1914: Some aspects of travel. *Geographical Journal*, 43(4), 365–77.

Koettlitz, R. 1900: Notes on geology and anthropology [to a journey through Abyssinia to the Nile]. *Geographical Journal*, 15(3), 264–72.

Kuklick, H. 1991: *The Savage Within: The Social History of British Anthropology, 1885–1945*. Cambridge: Cambridge University Press.

Lee, J. M. 1967: *Colonial Development and Good Government: A Study of the Ideas Expressed by the British Official Classes in Planning De-colonization 1939–1964*. Oxford: Clarendon Press.

Livingstone, D. N. 1992: *The Geographical Tradition*. Oxford: Blackwell.

Livingstone, D. N. 2002: Race, space, and moral climatology: notes towards a genealogy. *Journal of Historical Geography*, 28, 159–80.

Lochhead, E. N. 1984: The Royal Scottish Geographical Society: the setting and sources of its success. *Scottish Geographical Magazine*, 100(2), 69–78.

Louis, Wm. R. 1999: Introduction. In R. W. Winks (ed.), *The Oxford History of the British Empire*, vol. 5: *Historiography*. Oxford: Oxford University Press, 1–42.

Lycett, A. 1999: *Rudyard Kipling*. London: Weidenfeld & Nicolson.

MacKenzie, J. M. 1995: The provincial geographical societies in Britain, 1884–1914. In M. Bell, R. A. Butlin and M. J. Heffernan (eds), *Geography and Imperialism 1820–1940*, Manchester: Manchester University Press, 93–124.

Mackinder, H. J. 1900: A journey to the summit of Mount Kenya, British East Africa. *Geographical Journal*, 15(5), 453–86.

Mackinder, H. 1904: The geographical pivot of history. *Geographical Journal*, 23, 421–37.

Mackinder, H. 1919: *Democratic Ideals and Reality*. London: Constable & Co.

Madrell, A. M. C. 1997: Scientific discourse and the work of Marion Newbiggin. *Scottish Geographical Magazine*, 113(1), 33–41.

Marshall, P. J. 1996: 1918 to the 1960s: keeping the empire afloat. In P. J. Marshall (ed.), *The Cambridge Illustrated History of the British Empire*. Cambridge: Cambridge University Press, 80–107.

Mill, H. R. 1930: *The Record of the Royal Geographical Society 1830–1930*. London: The Royal Geographical Society.

Nash, C. 2000: Historical geographies of modernity. In B. Graham and C. Nash (eds), *Modern Historical Geographies*. Harlow: Pearson, 13–40.

Ogborn, M. 1998: *Spaces of Modernity: London's Geographies 1680–1780*. London: Guilford Press.

Ó Tuathail, G. 1996: Imperial incitement: Halford Mackinder, the British empire, and the writing of geographical sight. In G. Ó Tuathail (ed.), *Critical Geopolitics*. London: Routledge, 75–110.

Owen, N. 1999: Critics of empire in Britain. In J. M. Brown and Wm. R. Louis (eds), *The Oxford History of the British Empire*, vol. 4: *The Twentieth Century*. Oxford: Oxford University Press, 188–211.

Porter, B. 1996: *The Lion's Share: A Short History of the British Empire 1850–1995*. London: Longman.

Potter, R. B. and T. Unwin 1988: Developing areas research in British geography 1982–1987. *Area*, 20(2), 121–6.

Power, M. 2001: Geo-politics and the representation of Portugal's African colonial wars: examining the limits of 'Vietnam syndrome'. *Political Geography*, 20, 461–91.

Prescott, J. R. V. 1959: Nigeria's regional boundary problems. *Geographical Review*, 49, 485–505.

Rich, P. B. 1986: *Race and Empire in British Politics*. Cambridge: Cambridge University Press.

Robertson, G. S. 1900: Political geography and the British empire. *Geographical Journal*, 16(4), 447–57.

Ryan, J. 1997: *Picturing Empire: Photography and the Visualization of the British Empire*. London: Reaktion Books.

Sidaway, J. D. 2000: Postcolonial geographies: an exploratory essay. *Progress in Human Geography*, 24(4), 591–612.

Slater, D. 1997: Geopolitical imaginations across the North–South divide: issues of difference, development and power. *Political Geography*, 16(8), 631–53.

Smith, N. 1994: Shaking loose the colonies: Isaiah Bowman and the 'decolonization' of the British empire. In A. Godlewska and N. Smith (eds), *Geography and Empire*. London: Blackwell, 270–99.

Smith, N. 1999: The lost geography of the American century. *Scottish Geographical Journal*, 115(1), 1–18.

Spate, O. H. K. 1947: The partition of the Punjab and of Bengal. *Geographical Journal*, CX, 201–22.

Spate, O. H. K. 1954: *India and Pakistan: A General and Regional Geography*. London: Methuen.

Steel, R. W. 1964: Geographers and the tropics. In R. W. Steel and R. M. Prothero (eds), *Geographers and the Tropics: Liverpool Essays*. London: Longman, 1–29.

Steel, R. W. (ed.) 1987: *British Geography 1918–1945*. Cambridge: Cambridge University Press.

Watts, M. 1993: The geography of post-colonial Africa: space, place and development in Sub-Saharan Africa (1960–93). *Singapore Journal of Tropical Geography*, 14(2), 173–90.

Worthington, E. B. 1938: *Science in Africa: A Review of Scientific Research Relating to Tropical and Southern Africa*. London: Oxford University Press.

Younghusband, F. 1919: Address at the Anniversary Meeting. *Geographical Journal*, 54(1), 1–12.

Afterword: Emblematic Landscapes of the British Modern

David Matless, Brian Short and David Gilbert

While this volume sits within an emerging academic literature on twentieth-century Britain, it also draws on and seeks to feed wider cultural commentary concerning landscape, the built environment, and British culture, economy and society. Just as we should be aware of the cultural and political role played by academic geography through the twentieth century, so we might reflect on our own situation in offering this collection of essays. In this Afterword we consider how the processes of looking back on the twentieth century might themselves be reshaped by a geographical sensibility, a sensibility which has recently been applied to historically broader processes of British imperial and post-colonial memory and commemoration (Nash 2000). We have set out in the introduction to this volume the intellectual case for thinking geographically about the British modern, but will end here by reflecting on how the remembrance of modern things past might be informed by the geographical perspective presented in this collection.

Here again we enter a rapidly developing historiographic field, as bodies such as English Heritage and the Twentieth Century Society seek to identify and preserve built structures deemed to be of distinctive and characteristic value from the past hundred years. The language of preservation has been extended to the modern, in a manner which for some denotes paradox if not straightforward contradiction: is it not absurd to preserve structures which themselves gave no thought to the past? Of course modernism in architecture as elsewhere seldom rejected the past *per se*; rather, it worked with different narratives of it, constructed different traditions from which to oppose traditionalism. It is a narrow view of the modern which labels it as anti-historical, yet nevertheless there seems an irony in preserving structures which were built to denote confidence in the future, and whose vulnerability may suggest that such confidence was misplaced. A form of postmodern melancholic nostalgia for past confidence may accrue around the modern tower block, city centre, military installation, or power station, though on occasion, as with Bankside power station's transmutation into the Tate Modern, or the gentrification of high-rise blocks from 'problem' housing into luxury urban living – a process perhaps confirming that built structures do not determine social effects and that condemnations of buildings are often also judgements on people – preservation and reuse come into celebratory alliance. Whatever the motives and consequences, it remains the case that when such structures as Owen Williams's bridges over the newly created M1, discussed in Merriman's chapter here and featured on our cover, become possible targets for listing, the temporal and spatial values implied by making a preserved object for valued contemplation out of something designed to be passed at modern speed become complex to say the least (Glancey 1992; Saint 1992; cf. Taylor et al. 1975). There may of course be parallel ironies in earlier phases of these structures' lives, for example the plaque beneath a bridge on the M1 at Slip End near Luton commemorating the opening of the motorway is legally viewable only to those who might happen to break down on the hard shoulder beside it. Anyone else stopping to take in this historic site for modern motoring would be liable to a caution, now as in the motorway's early years (Merriman 2001: 219; cf. Sinclair 2002).

The work of the Twentieth Century Society and English Heritage, while focused on specific structures, points to ways in which particular sites become emblematic of particular modernities. The forms of geographical reflection in this volume may prompt wider thoughts on the emblematic landscapes of the British modern, not necessarily with a view to their preservation, but in terms of establishing further sites through which to conduct geographies of modernity which themselves may contribute to the re-evaluation of twentieth-century Britain. How might historical geographers work the twentieth century in ways which alert present society to the distinctive landscapes of Britain's recent past? We can approach this question through a geography textbook from the time of the 1947 Central Office of Information publication *Something Done*, discussed in our introduction, and which shares some of the same modern values, as well as echoing Mackinder's tradition of national geographic understanding. *Modern Geography. Book II: The British Isles* by Miss D. M. Preece, senior geography mistress at Crewe Grammar School, and Mr H. R. B. Wood, Director of Education for Wallasey, first published in 1939 and into its tenth edition by 1959, covers the country by theme and region in standard fashion. If its tone, unlike that of *Something Done*, is deliberately neutral, carefully non-judgemental, it remains expressive of a scheme of geographical value promoting careful and ordered regional development, and its coverage suggests sites worthy of historical enquiry. A vertiginous photograph looks 'down the pipe lines to Lochaber and the aluminium works', registering a scene of Highland regional development through a modern energy environment of hydro-electric power (Preece and Wood 1959: 209). Southampton Docks feature, with liners and cranes handling global trade: 'Note the storage sheds, some of which are specially equipped to handle fruit' (1959: 176). The ICI works at Billingham, coal mines and steelworks, Oxford University and the Cowley car works, Kent oast houses and Crawley New Town make up a modern British economic and social geography. All would warrant detailed historical geographical enquiry.

The strategy of national surveys alighting upon sites of significance is not confined to geographical textbooks. *Modern Geography* anticipates in textbook form Patrick Keiller's film of fifty years later, *Robinson in Space*, which offers another geography at the end of the century. The focus is on England rather than Britain, indeed Keiller's protagonist Robinson is commissioned 'to undertake a peripatetic study of the problem of England' (Keiller 1999: 6; Matless 2000). Keiller describes his film as concerned with 'the production of new space and the production of artefacts' (1999: 223). Armed with, but never overwhelmed by, quotations from such figures as Soja, Lefebvre and Massey, Keiller's Robinson moves an ironic, sardonic sensibility across the land, commenting on trading statistics, political intrigue, sexual mores, strange constructions, curious histories. Robinson is based in Reading, inspired by a mistaken reading of de Certeau: 'Reading is . . . a place constituted by a system of signs' (1999: 2). Up and down the Thames Valley and estuary, to Oxford and across to Milton Keynes, Bedford, Cambridge and Felixstowe, back to some south coast ports and on to Dorset and Bristol, through the West and East Midlands, over to Liverpool and on to Leeds, Hull, and Teesside, then back across to Blackpool and Cumbria and Hadrian's Wall to end in Newcastle with traffic over the Tyne and some local prehistoric rock carvings. The fixed camera and laconic intonation of fact and feeling through Keiller's roundabout excursion present a country fixed in limbo, with solid places made through trade, some rough and some sophisticated, and all going to make up a variegated 'problem of England'. The static gaze with overrunning words and figures is set up to displace rather than reinforce any clear, pinned-down, objectified understanding of a landscape; Robinson quotes Oscar Wilde to the effect that 'The true mystery of the world is the visible, not the invisible' (1999: 5). Odd things from history butt in, some still used, others whose function is unclear: the Cerne Abbas giant, Hull's white telephone boxes. Keiller works with a tradition of 'Condition of England' investigations running through the twentieth century via figures such as C. H. Masterman and J. B. Priestley, and is concerned especially with how, in a period supposedly dominated by a south-east-driven service economy, hidden spaces of manufacture and import/export continue to function. Out-of-the-way ports contribute to an economic geography of container transport and invisible earnings.

If the historical geographer contemplating a research agenda for mid-twentieth-century investigation could start with *Modern Geography*, those looking to conduct historical geographies of the immediate past could travel with Robinson. Such projects of national coverage themselves warrant attention as a form of commentary both culturally adjacent to historical geography and deserving of its analytical attention. A culturally inflected historical geography will look not only to key sites of British modernity but to the commentaries which helped shape them. So at the beginning of the twentieth century one might set Mackinder's *Britain and the British Seas* (1902) alongside Edward Carpenter's extraordinary Whitmanesque rendition of Britain in his 1883 *Towards Democracy*, issued in its complete edition in 1905, where, as in much left-wing commentary of the time, notably that of William Morris and Robert Blatchford, national landscape works within a romantic socialist internationalism, land and nation evoked as a means to critique the current state of both (Ward 1998). The key historical geographical move here is not only to connect Carpenter and Keiller as cultural commentators, but to place Mackinder and Preece and Wood in the mix alongside in ways which deepen our sense of what constitutes geographical commentary. Carpenter moves from universal and global matters (which he discussed in a fashion not unlike his contemporary French and Scots geographers Elisee Reclus and Patrick Geddes) to consider England and Britain, taking the aerial, visionary angle: 'England spreads like a map below me. I see the mud-flats of the Wash striped with water at low tide, the embankments grown with mugwort and sea-asters...' (Carpenter 1905: 54). And so on around coasts, Welsh and Irish mountains, rivers, cottages, cities, shipyards, palaces, industry, farming and ports ('I see the eternal systole and diastole of exports and imports through the United Kingdom' (1905: 57)), Carpenter moving ('I see', 'I explore', 'I hear') as a spirit assessing the wonder and horror of things:

> I see a great land poised as in a dream – waiting for the word by which it may live again.
> I see the stretched sleeping figure – waiting for the kiss and the re-awakening.
> I hear the bells pealing, and the crash of hammers, and see beautiful parks spread – as in toy show.
> I see a great land waiting for its own people to come and take possession of it. (1905: 58)

Carpenter seeks with his words to arouse the country towards democracy, and the following hundred years would in part be a story of the partial realization and continuing frustration of such landscaped geographical visions, whether Carpenter's nature socialism or a Fabian vision of state planning, visions themselves challenged by the continuing strength of those forces and values which Carpenter and his socialist colleagues and successors sought to undermine.

Having left shipyards and palaces, and before moving over manufacturing England, Carpenter offers a paragraph of agricultural geography, his vision moving over the land between local and regional scales:

> I see all over the land the beautiful centuries-grown villages and farmhouses nestling down among their trees; the dear old lanes and footpaths and the great clean highways connecting; the fields, every one to the people known by its own name, and hedgerows and little straggling copses, and village greens; I see the great sweeps of country, the rich wealds of Sussex and Kent, the orchards and deep lanes of Devon, the willow-haunted flats of Huntingdon, Cambridge and South Lincolnshire; Sherwood Forest and the New Forest, and the light pastures of the North and South Downs; the South and Midland and Eastern agricultural districts, the wild moorlands of the North and West, and the intermediate districts of coal and iron. (1905: 57)

Fifty years later Preece and Wood's *Modern Geography* registers its own modern agriculture, with maps of crop and animal distribution and aerial images of the Vale of Evesham and the Fens. We find where oats may be grown, look down photographically on the Fens near Boston, note the distribution of sugar beet. If some of Preece and Wood's modern geographic images of mid-twentieth-century Britain, notably those of manufacturing industry, document landscapes now lost, these pictures of farming show an emerging intensive agriculture, transformed from Carpenter's time, which has produced some of the most controversial late twentieth-century landscapes. The 'prairie' arable landscapes of East Anglia, for example, are as indicative of twentieth-century British modernity as the traces of open field farming may be of medieval historical geography, whether they are seen as signs of progress and efficiency or ecological deserts signalling the 'theft of the countryside' by those previously upheld as its custodians. Marion Shoard influentially introduced this 'Paradise Threatened' in 1980: 'England's countryside is not only one of the great treasures of the earth; it is also a vital part of our national identity.... Although few people realise it, the English landscape is under sentence of death.... The executioner... is the figure traditionally viewed as the custodian of the rural scene – the farmer' (Shoard 1980: 9). If Howkins's chapter in this volume explores primarily negative reactions to such sites, one could speculate whether, in a future where less intensive forms of agriculture might prevail, a prairie arable farm in East Anglia might become as rare and historically indicative as the open field village of Laxton in Nottinghamshire, that staple of twentieth-century historical geography curricula (Orwin and Orwin 1938).

Such a scenario might seem unlikely, but in former coalfield regions it is easier to imagine grass-covered spoil heaps becoming the equivalent of medieval ridge and furrow for the next millennium, bumps in the landscape indicating the trace and scale of a former way of life. One might also consider other lost or surviving sites contemporaneous with Preece and Wood's *Modern Geography* which, while not registered in a mid-twentieth-century geographical vision, say something about the era it sought to understand and shape. The 1950s milk-bars quizzically scrutinized by Richard Hoggart in *The Uses of Literacy*, or the miniature model villages constructed before and after the Second World War as rural tourist magnets or sideshows to the seaside resort, may have been overlooked by the mid-twentieth-century geographer as trivial or whimsical, but now appear equally suggestive of the economic and cultural geographies of the time. Milk-bars gained their academic chronicler in Hoggart, and may now be memorialized as much for their place at the origin of cultural studies as for the fun had within them. Here hung the 'juke-box boys', and here a cultural geography played out of young British bodies and minds taking on American style via a new consumerism, which for Hoggart signalled an erosion of local working-class cultural life. The placeless milk-bar did not match the local:

> Compared even with the pub around the corner, this is all a peculiarly thin and pallid form of dissipation, a sort of spiritual dry-rot amid the odour of boiled milk. Many of the customers – their clothes, their hair-styles, their facial expressions all indicate – are living to a large extent in a myth-world compounded of a few simple elements which they take to be those of American life. (Hoggart 1957: 248)

If Hoggart indicates one line of enquiry for a cultural historical geography of the 1950s, miniature villages would likewise illuminate the geography of the time. Sites such as Bekonscot near Beaconsfield (1929), Bourton in the Cotswolds (1937) and Pendon in Oxfordshire (1930s) were the forerunners of a series of seaside model villages constructed after 1945 in resorts such as Skegness, Southsea, Babbacombe and Southport. Merrivale model village, Great Yarmouth, opened in 1961, offered 'a typical section of the English countryside in miniature', a version of rural Englishness amid the seaside Englishness of Yarmouth, with farms, factories, villages, sports stadium, castle and the market town of Merritown: 'All in all, nothing has been omitted from this Lilliputian land of make-believe; it is the most authentic and realistic of models that delights young and old alike.' The simultaneous claim to authentic realism and make-believe indicates the space such sites could occupy in cultural life, at once evidently contrived yet signalling a supposed cultural truth about rural English life, traditions laid out in the middle of the modern seaside yet with their own modern elements: 'Merritown offers first-class shopping facilities for its inhabitants' (Anon. 1972). If, as Edwards shows in this volume, Simpson of Piccadilly displays a version of national identity, so too does Merritown's supermarket and stores. This particular pastoralism of ideal rural settlement – farm, village, market town – not unlike the contemporary visions conjured in the miniature televisual worlds of Trumptonshire, or the later children's collectable model village toy 'Sylvanian Families' (Houlton and Short 1995), would repay further study. The survival of such sites today might well exercise the Twentieth Century Society.

Ports, mines, factories, fields, villages miniature and full-size, power stations, whether hydro, coal or nuclear, all offer sites through which to access the geographies of mid-twentieth-century British modernity. Some remain for more or less easy field study, others would require an archaeological dig to find an outline. Moving across such sites in a particular time period we might find potential for a renewed version of the historical-geographical trope of cross-sectional analysis, taking a cut through the culture, polity and economy of a time. Equally, what is suggested here fits within a long-standing pattern of historical-geographical study of emblematic sites, or indeed processes whereby sites are made emblematic through study; the medieval open field village of Laxton, the deserted village of Wharram Percy, the drained fens of Cambridgeshire. Interest in the twentieth century is of course driven in part by the dramatic changes felt through the period, changes registered within this volume; in social legislation, in rural land use, in ethnic identity, in travel and home life. Transience will be an abiding theme of any historical geography of the twentieth century, as of earlier periods. If we abide in transience, however, interest in a recent century which has produced so many extant landscapes is also prompted by what might be termed the continuing weight of geography. The spaces through which twenty-first-century British lives are led derive from previous times, and most notably from the past hundred years, and any journey across Britain shows the usual mix of the newly made (whether proudly modern or imitating historical, even twentieth-century, styles) and ruined old (whether priories or factories) with large areas where things proceed in a built or cultivated environment whose form is stamped with the motives, economies and dreams of the twentieth century; not least in the suburbs, the modern agricultural landscape, in Forestry Commission lands, in post-war city centres, in New Towns, in nuclear stations, in national parks, on motorways. Some of these sites are felt to be becoming old, gaining period charm or lingering too long. Some seem much as the day they were constructed on green fields or bomb sites. Some retain novelty and/or notoriety, with nuclear sites novel and notorious in part for the fact that their traces can never disappear. The geographies of British modernity presented in this volume begin to account for such spaces, though they leave more unexplored.

REFERENCES

Anon. 1972: *Merrivale Model Village*. Norwich: Jarrold.
Carpenter, E. 1905[1883]: *Towards Democracy*. London: George Allen & Unwin.
Central Office of Information 1947: *Something Done: British Achievement 1945–47*. London: HMSO.
Glancey, J. 1992: A bridge too far? *The Independent Magazine*, 18 July, 24–31.
Hoggart, R. 1957: *The Uses of Literacy*. London: Chatto & Windus.
Houlton, D., and B. Short 1995: Sylvanian Families: the production and consumption of a rural community. *Journal of Rural Studies*, 11, 367–85.
Keiller, P. 1999: *Robinson in Space*. London: Reaktion Books.
Mackinder, H. 1902: *Britain and the British Seas*. Oxford: Clarendon Press.
Matless, D. 2000: The predicament of Englishness. *Scottish Geographical Journal*, 116(1), 79–86.
Merriman, P. 2001: M1: a cultural geography of an English motorway, 1946–1965. Ph.D. thesis, University of Nottingham.
Nash, C. 2000: Historical geographies of modernity. In B. Graham and C. Nash (eds), *Modern Historical Geographies*. Harlow: Prentice Hall, 13–40.
Orwin, C. S., and C. S. Orwin 1938: *The Open Fields*. Oxford: Clarendon Press.
Preece, D. M., and H. R. B. Wood 1959 [1939]: *Modern Geography. Book II: The British Isles*. London: University Tutorial Press.
Saint, A. 1992: *A Change of Heart: English Architecture Since the War, a Policy for Protection*. London: Royal Commission on the Historical Monuments of England/ English Heritage.
Shoard, M. 1980: *The Theft of the Countryside*. London: Temple Smith.
Sinclair, I. 2002: *London Orbital*. London: Granta.
Taylor, R., M. Cox and I. Dickins 1975: *Britain's Planning Heritage*. London: Croom Helm.
Ward, P. 1998: *Red Flag and Union Jack: Englishness, Patriotism and the British Left, 1881–1924*. London: Royal Historical Society.

Index

KILLER CHRISTMAS

An Emma Wild Mystery Book 1

HARPER LIN

Harper Lin Books

I've had people faint in my presence before, but never had one to drop dead at my feet—literally. The girl I was chatting with at the cafe suddenly began to choke and shake. Her drink slipped from her hand and she collapsed onto the floor. Her body convulsed and her eyes rolled to the back of her head. The spilt hot chocolate made it look as if she were lying in a pool of brown blood.

I knew the drink had been poisoned. I knew it in my gut when I bent down to inspect her body and the drink on the floor. And I also knew that I was somehow responsible for what happened. She died because of me. But I didn't admit any of this to myself until later.

At the time, however, I only thought that perhaps it was a good idea to stay away from hot chocolate.

Maybe I should start at the beginning. Or should I say the end? A story can begin with heartbreak or end with one, but I was sure not willing to let both happen.

In my instance, the story began with heartbreak and I was not willing to let it end until a happy ending was reached.

My name is Emma Wild. Yup, that Emma Wild. The crooner with the two Grammy award-winning albums. Jazz, blue-eyed soul, alternative and main-stream pop are some of the ways they would try to categorize my music.

How would I categorize myself? An incurable romantic. And a singer.

I wasn't always that girl on MTV singing about heartbreak in a husky alto voice choked full of tears. I didn't always wear body-hugging dresses and my red hair in glamorous waves styled over one eye. I grew up in Hartfield, Ontario, a town about an hour and a half from Toronto. It was a charming town and my family still lived there, but at eighteen years old, I tailed on out of there, forgoing college to pursue my singing career. For two years, I busted my ass singing at every open mic in New York City. I knocked on every door and pushed demo tapes into the hands of anyone who was connected to anyone until finally somebody gave me a chance. Then it became a blur from there.

My record company was behind me every step of the way—at least in the looks department. I was a control freak in the studio, so I let them play with my hair and makeup and put me in designer clothes and on the cover of magazines while I picked which producers, musicians and video directors I wanted to work with.

Along with all that came the world tours—Paris, Tokyo, Melbourne, you name it. TV appearances, award shows, press conferences, and movie premieres consumed my years. A "thank you" and a smile here, a funny quip there, a twirl to show off who I was wearing—I could do all that stuff in my sleep. Literally. Sometimes I was so tired from all the work and travel that I would give an interview while I was half asleep. Still, I had fun with it; playing the fame game came pretty easily to me because I didn't take it seriously.

The only thing I took seriously was my work, the music. I'd been writing my own songs since I learned how to write my name. In high school, I played clarinet in the school band...nerd alert. I sang at every event where somebody allowed me to take hold of a mic. I'd even go to poetry slams and sing with a guitar instead.

During my childhood and teen years in Hartfield, every winter I'd sing at the Christmas concert at the town square, and every summer I'd sing at the food festival. Then there was everything in between: talent shows, private parties, baseball games. Whoever needed a singer would only need to speed dial my mom.

So it didn't surprise my friends and family that I would make it big with all the ambition and the steely determination that I possessed. Due to my busy schedule, their relationship with me was pretty one-sided most of the time. Sure, I called them whenever I could, and I would fly them into New York, where I lived, but on a day-to-day basis they learned about what was happening in my life through the tabloids. Especially in recent years when I started dating someone more famous than I was.

In general, my private life was pretty tame up until I met him. All I did was work. I'd ignore all the stories in the papers, magazines and blogs written about me to keep my sanity and self-esteem intact.

After dating a few industry types, I'd sworn celebrities off, bored with their massive egos and self-entitlement. Then along came Nicolas Doyle. Yes, that Nick Doyle. The movie star who had a bad habit of dating supermodels until he met me. Me, a five foot two redhead with pale skin and freckles—a little ball of fire, as the journalists would sometimes call me. I didn't think I was his type; I wasn't under contract with Victoria's Secret as one of their Angels. But once Nick has his eyes set on something, or someone, he usually got what he wanted.

Soon I started showing up at his movie premieres, separately. We tried our best to keep our private lives private, but it didn't take long for the tabloids to put two and two together, and soon I was being photographed in my sweatpants during my daily morning runs in Central Park.

When I first ran into him at the Vanity Fair party, I was joking around with a bunch of joke writers. They were a pretty funny pack. They kept teasing me for how bad my jokes were and kept trying to make even worse jokes to one-up me. Nick joined in, seemingly out of nowhere, with an extra glass of champagne and a witty remark of his own (something about horse butts—don't ask), and I downed the glass to keep myself from shaking.

Back then, he wasn't Nick. He was the Nicolas Doyle that the public knew. The piercing blue eyes that burned through movie screens, the mischievous grin, the raw talent that allowed him to disappear into any role—I'd grown up with him from baby-faced TV star to strong-jawed leading man.

I was totally starstruck at first, but I fell in love with him after I knew him. I loved that he got involved with all sorts of causes. All the charity work that he did during his time off was not a publicity act, I came to find out. It was what made him irresistible. Underneath all the Hollywood hoopla, he was a caring guy. When he had passion for something, he threw everything into it, whether it was acting, saving extinct pandas, or brightening up the lives of children born with cleft palates. So when he didn't want to get married after four years of being together, I knew that something was wrong.

When a guy could take or leave you, it wasn't a good sign. I'd been down that road before, got the T-shirt and didn't want to go back there again. It was in my best interest to move on, even if it was the hardest thing I had to do—maybe the second hardest.

I packed all my stuff from his New York pent-house and stayed holed up for a week in a hotel, churning out song after song on this little hotel notepad. I couldn't think about finding another apartment yet, but I knew I had to eventually. I didn't know if I even wanted to stay in the city, where

I'd be constantly paranoid about running into him or his millionaire friends.

Luckily, Christmas was coming up and I had the excuse to go back to Canada to spend the holidays with my family for a while, so I booked the next flight to Toronto, then hired a driver to take me to my hometown.

It was in the car that I began to panic about something else. I hadn't been back in Hartfield for at least two years.

It wasn't until I was getting close to the town when I remembered that I stayed away from Hartfield as much as I could for good reasons. One, to be exact. The cause of the first story of my life. The one that ended in heartbreak.

CHAPTER TWO

I haven't celebrated Christmas in Hartfield for five years. Usually around this time, I'd be touring or doing promotional work somewhere in the world, or spending the holidays with Nick's family.

I never thought that staying away from Hartfield was a conscious decision, but now I saw that maybe it was. I had flown my family to Mount Tremblant in Quebec for most of those winters and we'd spent time together there. I'd meet them on neutral grounds some weekends, mostly in New York, where I lived and they never got tired of visiting. They would also fly to my concerts all the time, which could be in any part of the world.

Since Mirabelle got pregnant five months ago, it'd

been more of a challenge for her to travel in recent months, so I was eager to see her most of all.

I did love Hartfield at Christmas. It was so lovely during this time of year. The Christmas Market was probably all set up at the town square. Performers would sing, play music or put on shows most evenings. There would be a little more pep in the townspeople's steps, although the stray from their usual lax attitude could also be from the stress of shopping and finding the right gifts for their loved ones.

And of course, the best place in the world was sitting in front of the fireplace at my old house, nursing a cup of my dad's eggnog and chatting with my family. The lovely smell of something baking in the oven would waft in whenever somebody passed through the door from the kitchen to the living room, and Christmas muzak would set the mood from the vinyl player in the corner.

In a word, Hartfield was cozy this time of year.

All I had to do was put my sour memories in the gutter and focus on having a joyous time with my family.

But as I sat in the back of the car that was driving into Hartfield from the airport, I suddenly started to cry.

The stress was getting to me. It had been an

emotional week. I'd been with Nick for so long and I was madly in love with him. I still was, but my pride gave me the strength to walk away. I knew if I didn't get out now, I'd be losing the best years of my life to someone with commitment phobia.

The driver was kind enough not to notice the sobbing, and even turned on the radio a little louder out of consideration.

After a few songs, one of mine came on. "Falling Into Pieces" was a smooth jazz song that I'd written last year about how much I loved Nick. That made me sob even harder. It really hurt to love a man who didn't return your love the way you wanted him to. But I was too stubborn to accept anything less. I'd written a handful of new songs about that recently.

When we reached Hartfield, my tears had more or less dried. I looked out the windows at the fresh white snow blanketing over the houses and the stores. Christmas decorations were everywhere, with lights strung up from lamppost to lamppost. The sight of my old hometown cheered me up; a smile began to crack on my face.

I instantly regretted being too chicken to come back every Christmas. This was the best place to be this time of year. Plus, nobody was here to hound me with questions about my personal life. No paparazzi here and the townspeople who did recognize me left me alone. This was Canada after all, where no one was overly impressed by anything. I was used to being mobbed by fans and paparazzi in big cities around the world. Getting your picture taken when you have no makeup on, wearing sweats, and just want a cup of coffee from down the block did very little for your self-esteem when you were torn to shreds on some gossip blog or Page Six the next day. Hartfield could've been my sanctuary.

When the driver pulled up in front of my parents' house, my breathing was no longer punctuated by uncontrollable sobbing. The driver helped me with my luggage and I slipped him a huge tip. Any acts of kindness from strangers were magnified tenfold and I was beyond grateful.

"Thank you so much, Carl," I said to the driver.

"You're welcome," he said with a wink. "Take care of yourself. Oh, and my wife is a big fan."

"Oh really? What's her name?"

"Sadie," he replied.

I reached into my purse and pulled out a copy of my third CD. It wasn't released yet, but I'd brought some extra copies for family and friends.

I signed it and gave it to Carl.

"Here."

A toothy grin appeared on his face and I couldn't help but smile back.

"Merry Christmas," he said.

"Merry Christmas!"

When he drove off, I took in the crisp air. But my moment of appreciation was cut short.

A woman stared at me from across the street. She looked like she was either squinting in my direction or giving me a dirty look. She was dressed in an over-sized wool sweater, her strawberry blonde hair in a messy ponytail. She looked familiar.

"Emma Wild," she called.

At the sound of her shrill voice, I recognized her instantly. Kendra Kane. I'd call her my nemesis if the term didn't make me laugh. We were friends in grade school until her bossy nature conflicted with my independent one. She was the mean girl of the school. Maybe she was every girl's nemesis.

I couldn't stand the way she was always so competitive, trying to outdo me, or anyone really, every chance she got. I couldn't stand girls who brought other girls down to lift themselves up, and I largely ignored her from middle school on. That was when I started hanging out with Jennifer and Cassandra, who still remained my two BFFs, although they were both scattered around the country now.

Unlike most of the other girls who moved to big cities as soon as they graduated high school, Kendra was still around.

"Hi, Kendra," I said, putting on a smile. "Good to see you."

She only smirked in response. Nonetheless, she crossed the street to where I was standing. I braced for the worst but hoped for the best. We were both pushing thirty, too old for any more teenage cattiness.

"Giving the fans what they want, I see?"

She must've seen me autograph the CD. There was a certain awkwardness about being famous that I didn't like to think about unless someone brought it to my attention.

"So, what have you been up to?" I asked. "It's been so long."

She waited before answering. Kendra had a habit of making me feel uneasy. She stared at me, but there was a blankness in her cool green eyes, a deadening chill that made me feel as if she wasn't all there. She had a way of exercising her power this way.

"A lot has happened," she said. "I got married and had a baby boy."

"That's great! What's your boy's name?"

"Blake, Junior."

I was genuinely happy for her. Maybe she had changed. People did grow up, after all.

"Wow. So you live across the street now?"

"Just recently. Couldn't pass up the opportunity."

"That's great. I didn't know that. It's a lovely house."

"Yeah, well, I'm sure you've been busy singing your songs."

She said it as if it were a dirty word. I couldn't let her get to me and maintained the smile on my face. This was nothing. If I had a dime every time I smiled pleasantly at someone who was mean to me...

"Yup," I replied, smile still plastered on my face. "My third album's coming out, so I'm happy about that."

"Congratulations," she said flatly. "I'll be sure to buy a copy."

"No need." I reached into my bag.

When I handed my CD to her, she shook her head.

"I want to buy it," she said. "You know, to support you. Maybe you should give this to Sterling. I'm sure he'll appreciate it."

At the sound of his name, my smile faltered. My cheeks burned and suddenly I wanted to throw my CD at her face. She always knew which buttons to push. Kendra smirked again just as my mom opened the door.

"Emma!" Mom said. "Hello, Kendra."

Kendra gave a half-hearted wave.

"Lovely catching up," she said before she walked back home.

Mom ran down and hugged me so tightly that I couldn't breathe.

"Hey, kiddo." Dad helped me with my suitcase.

"Kendra lives across the street now?" I said when we were inside.

"Don't sound too excited," Dad said.

I supposed there was a groan in my voice.

"The poor girl's been through a lot," Mom said. "She needs your support."

"I very much doubt that," I said. "But what do you mean?"

"Her husband Blake died last summer."

"Died? She didn't mention that."

"Yes. She was head over heels in love with him, but one day she came home and found him dead."

"How?"

"We don't know the details," said Mom. "But I did hear from somebody in my knitting circle that he had a stroke or something."

"God, I didn't know."

I felt bad now for wanting to throw that CD in her face. She had her reasons for being a miserable person. She wasn't angry at me. She was angry about life.

"Poor thing," Mom continued. "She'd been a

single mother raising that little boy of hers. They lived in a house down on Lakeshire, but after the funeral, she sold the house and moved here when Margot passed away. It's closer to the elementary school, and I'm sure she didn't want to live in a house full of memories of her husband."

"She didn't want anybody at the funeral," Dad added. "I don't even think she held a funeral."

"I should do something nice for her," I said. "Maybe we can invite her over for dinner one day."

"It's hard to reach out to her," Dad said. "She's closed off, doesn't like to talk to people and never wants to go to any social gatherings. Neighbors keep trying to go to her house and offering her all kinds of food and advice, but she's not having it. The death is still fresh, so I think she's still grieving."

"I just can't imagine what she's going through," I said.

I supposed everyone had personal tragedies. Mine suddenly didn't feel so life shattering.

CHAPTER THREE

There was something utterly bizarre about being back in my childhood room. It felt as if I were stepping into a time capsule. Everything was tiny. My little twin bed still had the flowered sheets, the candy-striped wallpaper was a little faded, and the little desk was still by the window where I used to do my homework. I was glad that my parents didn't want to renovate anything; I was glad to find everything still the same.

Except when I closed the door. That was where a huge poster of Nick still hung. He was in his teens. His dirty blond hair was cut in a mushroom style that had been all the rage back then, and he was smiling and petting a golden retriever. This was the poster for that dog movie that I loved when I was young. So basically I'd had a crush on him since I was fourteen. How depressing.

This was going to be a hard breakup to get over, even harder than the first one, which I still didn't want to think about. One heartbreak at a time was plenty for me.

I took down the poster slowly, careful not to rip it. I couldn't bring myself to throw it away, and I ended up rolling the poster up and hiding it in the back of my dusty old closet.

Other than that small, painful reminder, I was happy to be back in my old room. I got under the covers of my bed and slept because I was so tired from the early flight.

After an hour, a bang on my door woke me up.

"Kiddo?" Dad called. "Lunch is ready. Wanna eat with us?"

"Yes," I replied. "I thought you'd never ask."

At the offer of food, I practically jumped out of my bed.

There was nothing like Dad's comfort food to get you out of a funk. Sure, I'd eaten in some of the finest restaurants around the world, but nothing compared to Dad's beef stew and crab cakes.

"You didn't mention on the phone whether Nick was coming for Christmas?" Mom inquired.

I tried not to cringe at the mention of his name. For the past year, Mom had been waiting for Nick to propose. If I had to be honest, I was too.

So I had purposely forgotten to tell them. The plan for me to come home for Christmas was last minute, and of course they were thrilled, but I didn't exactly want them feeling sorry for me.

"His schedule is crazy," I said. "I'm not sure. He has this new movie coming out..."

"I can imagine," said Mom. "But I hope he can make it. He's never been to our town. It would be nice to show him around your old haunts, you know."

"Of course. It's too bad."

My family had met Nick on several occasions and they adored him. Dad was a big fan of *Alive or Dead*, the only action film Nick had ever been in. Mom was still bewildered that one of the biggest movie stars in the world was dating me.

I changed the subject.

"Is Mirabelle at work?"

"Should be," Dad said. "She always is this time of day."

"Great. I'll go and swing by the cafe after."

I couldn't wait to see my big sister. She lived only one block away and she owned the Chocoholic Cafe, the best cafe in town. They had the best coffee. I was the one who'd helped her pick all the beans before she opened the store after all—and yummy chocolate desserts to go with it. Their specialty was their organic hot chocolate. Of course, since I loved caffeine and chocolate equally, I made her invent a drink that combined the two, the hot chocolate latte.

On Samford Street, where all the best shops were, the Chocoholic Cafe was buzzing with locals. I couldn't wait to get my hands on all their fresh chocolate cakes, croissants and cupcakes during my stay.

When I arrived, Mirabelle wasn't behind the counter. There were two baristas making the drinks, Kate and Michelle. They looked busy, but I approached them to ask where my sister was over the sounds of coffee beans being ground to death and a latte machine making the sounds of a dying cat.

"She's out running a few errands," said Michelle.

"Great, I'll just wait for her then."

"Want me to make you something?"

"Thanks, but I'll wait in line so we don't get chewed out. You can't piss off people who are demanding caffeine or chocolate, or even worse —both."

Kate and Michelle laughed.

"Amen," said Kate.

While in line, I checked my phone to catch up on the news. Usually I avoided the entertainment section like the plague, but a fresh headline caught my eye.

Nick Doyle Dumps Emma Wild for New Victoria's Secret Angel.

Seriously?!

Nick was pictured smiling next to a laughing blonde. He was wearing a tux and the model was in a slinky red dress that showed off ample cleavage. I was used to wild rumors, but my jaw dropped at this one. Before me, Nick did date half of those angels. Now he was onto this child who looked barely twenty years old when I only moved out a week ago?

Nick Doyle stepped out in a dapper Armani tux yesterday night for the Sick Children's Charity Benefit. Sources say that his longtime girlfriend, singer Emma Wild, moved out from their Soho apartment recently and he is currently single and already mingling. Witnesses caught him in deep conversation with new Victoria's Secret Angel Tara Amberstone. They couldn't pull their gaze away from each other all evening. This isn't the first time Nick Doyle has been smitten with an Angel. In the past, he has dated...

I forced myself to stop there. I believed this article. Completely. Moving on this quickly was something that Nick would do. He'd done it in the past to ex-girlfriends. In fact, he'd only broken up with his supermodel ex-girlfriend a couple of weeks before he began flirting with me. Why did I think I would be any different?

Still, the article stung. I wanted to lock myself in a room somewhere and break down. This was so humiliating. Now everyone was going to know. My parents would find out eventually. Everyone would feel sorry for me.

Luckily, most of Hartfield's population were my parents' age. They were more into crossword puzzles than trashy celebrity gossip. I hoped.

Taking a deep breath, I closed my phone and put it back in my purse. I concentrated on the chocolate pound cake sitting above the counter next to a tray of gingerbread cookies.

Inhale. Exhale. Inhale. Exhale.

The guy in front of me turned around and gave me a weird look.

Yup, I was breathing too hard.

I looked at the colorful cupcakes, the muffins, the oversized cookies and breathed in the sweet aroma of the cafe. Chocolate would cure me. Caffeine would cure me.

I wasn't going to cry.

It was finally my turn to order. I could put on my pageant smile again.

The cashier was a lanky hipster type in his late teens or early twenties. With shaggy dyed black hair over his eyes, a ring dangling out of one nose, and heavy eyeliner, he looked out of place in this quaint cafe.

His name tag said "Cal."

"Hi, what can I get you—Emma Wild?"

His hazel eyes flashed with recognition when they met mine.

"That's me."

I was surprised that he knew who I was. He seemed the type to listen to heavy metal or punk.

"I...know you," he said somewhat awkwardly. "Mirabelle's sister."

"Oh, right."

That was why he knew who I was. Of course.

After placing my order, I stood with a bunch of people by the counter to wait. Since Cal had already given me my piece of chocolate pound cake, I started nibbling on it from the bag. I had also bought half a dozen chocolate chip cookies, two chocolate biscotti, a chocolate croissant, and three cupcakes. If they'd sold ice cream, I would've bought that too. I reminded myself to buy ice cream from the supermarket on the way home. There was none left in our freezer, if you could believe. And I wasn't going to share any of this stuff. It was breakup food.

I tried to remind myself that I didn't have it so bad. Nobody was immune to heartbreak in some form. Look at Kendra. I couldn't imagine losing a husband so young. This was nothing. After I had a good cry and stuffed myself full of sickeningly sweet desserts, I'd be fine.

I was sick of being heartbroken. And singing about it. My last two albums hammered this subject to death. The third one coming out was full of songs about love and loss too.

I'd been feeling this way since I was old enough to go out on a date. It was time to finally hang up this shtick and be in a relationship where I was treated like a queen. First, I'd make an effort to write some songs that didn't have to do with love and relationships.

Maybe for my next album, I would do something different. Maybe some brainless pop numbers, stuff that people would feel happy listening to. Or maybe I'll just stay single, put my career on hiatus and do some volunteer work halfway around the world for a while.

But who was I kidding? I loved music. I couldn't go a day without some new tune taking over my head. If this wasn't true love, I didn't know what was.

As I stood there thinking about all the career

routes I could take with my musical direction, one of the baristas called out my name. My tall cup of hot chocolate latte was steaming and ready for me.

But before I could reach for it, a blood-curdling scream came from behind me.

CHAPTER FOUR

The scream almost made me drop my box of desserts and my bagged piece of pound cake. I whipped around and faced a crazy Cheshire grin on a woman around my age. She had a blonde bob, wore a red tuque, and simply could not stop jumping and clapping her hands.

"Oh my God!" She stepped forward, invading my personal space by a nose hair. "Emma Wild! I heard that you're from here, but I can't believe that I'm actually looking at you in person! You're so thin! I mean, in a good way. And your skin. What's your secret?"

The other patrons looked in our direction and I blushed.

"Heeey." Now that I was in the presence of a fan,

I had to be on. "Um, I don't know. I try to sleep and eat well."

"I am such a big fan," she gushed. "Your last album is still on constant replay on my stereo. 'The Killer in Me' is just a phenomenal song. I was just singing that in the shower this morning! What a coincidence! Seriously though, you're really an incredible songwriter. Look at me, I'm totally blabbering, but I can't help it. You're, like, such an inspiration to me."

"Thank you so much." Pageant grin, check. Modesty, check. Incessant nodding, check.

As much as I appreciated meeting fans who enjoyed my work, I tried not to let their constant compliments inflate my head. I had to learn to be immune to what people said about me so that opinions, good and bad, didn't affect who I was.

"Soy latte?" Michelle, the barista, called out.

My fan signalled to Michelle that it was hers, but she was rooted in place before me.

"I can't believe you're here." She was still jumping. "I know I'm really not acting cool right now, but I love you so much. I can't believe you've been standing in line in front of me this whole time and I didn't know it. How long are you back?"

"Just for the holidays," I replied.

"Maybe we can hang out, you know? I always

thought that we could be friends? I mean, I see you in magazines, and I think we have similar style."

"Oh?" I laughed awkwardly.

I really didn't know who this woman was, and I quickly tried to figure out a way to get out of this. If she lived in this town, she might want to hang out all the time.

"Well, I don't know," I joked. "You'll have to pass my test. What's your favorite TV show?"

"*The Voice!*" she exclaimed. "I think you should be a judge on that show. You're so much better than Shakira or Christina."

"Oh, that's not true. Those girls are super talented. I've met both of them and they're very nice. So you're from around here?"

"I live in Sanford actually, but I drive back home from work through this town and I absolutely love this cafe."

She didn't live in Hartfield. That was a relief to hear.

"What did you get?" I asked.

"The organic soy latte is seriously good. Have you tried it? What did you get?"

"I have. I love everything here, but my fave has got to be the chocolate latte. Have you tried it?

"No, but I will now that I know that Emma Wild likes it."

Even though I was waiting for Mirabelle, it was probably best to leave. I reached for my cup on the counter and moved to the area where all the lids were. She followed me with her drink and put a lid on her cup quickly. I hoped she didn't want to leave with me. Before I could give an excuse, she asked for an autograph.

"Sure." I prayed that she wouldn't ask for my number or to hang out again.

She began rummaging through her purse until she found a notepad and a pen. I put my cup on the counter to give her the autograph as she drank from her cup. When I gave the notepad back to her, I saw that her face had scrunched up.

"This drink tastes really weird. I think—"

She clutched her chest, yelping a little. Then she began to spasm. Her cup slipped from her fingers and her body fell after it.

"Oh my God!" I shrieked.

The crowd at the cafe circled around her.

"Call the ambulance," Michelle said to Cal.

"Is she having a stroke?" a customer asked.

"Is there a doctor here?" I shouted out to the sea of shocked faces.

A woman wearing blue scrubs under her winter coat stepped forward. "I'm not a doctor, but I am a nurse."

I moved out of the way as she crouched down beside the fallen fan.

"Everybody clear out," said Michelle. "Please give us some space."

The customers began to go reluctantly.

"Please!" Michelle urged again. "The ambulance is coming and we need to make room."

"I don't feel a pulse," the nurse said.

"The ambulance is on its way," Cal said. He stepped out from behind the counter, face beet red.

"What do you mean there's no pulse?" I exclaimed to the nurse.

I looked down in horror at the girl I'd just been chatting with. She'd looked so flushed and vibrant only moments earlier and now...

"She's...?"

"Dead," the nurse said. "I think she's dead."

A hush went through the crowd. Cal, Michelle and Kate all looked horrified.

I looked down again at the girl who had just claimed to be my biggest fan. She was now a corpse.

CHAPTER FIVE

I couldn't understand what had happened. One minute she was alive and full of energy and the next minute she fell to the ground. It didn't make any sense. I kneeled down and looked at her in that brown pool of hot liquid. Blonde hair covered her face and one of her arms was crossed over her stomach. She had mentioned something about the drink tasting strange before she fell...

"The ambulance is here."

I looked up and saw Cal. His nervous red face loomed over me.

"Please leave," he said.

I stood up and walked out the door, numbed by the whole incident. The chill of the winter cold numbed me further and I moved out of the way as the paramedics went in.

But what could they do if she was already dead?

The coffee shop crowd had more or less stayed put outside. I stood with them, watching and waiting.

"This is horrifying," an old woman beside me remarked to her friend.

"Poor thing," the other lady replied. "I hope she's all right. Maybe she just fainted."

All around me the townspeople talked amongst themselves. A couple of pedestrians stopped and asked what was going on. The two ladies tried to explain, but they ultimately said that they didn't know.

"Miss?" The first lady turned to me. "Do you know what happened? Did she have a stroke?"

I couldn't tell them that she was dead. "I don't know the details."

"I heard that you're famous or something?" The second woman asked. "I'm sorry, I don't recognize you, but the young lady did. Were you talking before she fell?"

"Well, yes, we were chatting when we were getting our drink—"

Then it hit me. How could I miss this detail? She had ordered a soy latte. The cup that she drank from and then spilled was dark brown. She had taken the wrong drink. Maybe she'd been allergic to dairy and that was why she requested soy. What if she had been seriously allergic to dairy?

I walked back to the scene to tell the paramedics. The police were also inside, and as I went in, I noticed Cal was talking to one of the two policemen and all of them turned to me.

"Miss?" said the cop with a scruffy beard. "I heard you were talking to the victim before she fell."

"Yes," I said. "Is she really dead?"

The taller cop nodded grimly. "Unfortunately she is. We don't know why, but we'd like to know what took place between you two so we can find out more."

I repeated what we talked about, embarrassed to mention that I was a celebrity, and that I was giving her an autograph before she fell.

"Oh, I think I recognize you," said the tall cop. "You sang that song, 'Cornflower Blues.'"

He looked at me with more interest.

"That's me," I said, and then quickly changed the subject. I told them the victim had taken the wrong drink.

Her body had already been removed from the floor, but the cup was still lying there. A name was written on the cup: Emma.

I looked around for my cup, which was still on the counter where I left it. It also said Emma.

"That woman—her name was also Emma?" I asked the cops.

"Yes. Emma Chobsky."

I looked inside my drink and smelled it. Soy. It was the other Emma's soy latte.

"Our drinks got mixed up," I said. "She took my chocolate latte. Her drink was a soy latte. It could be that she had a severe allergy to dairy. I feel horrible."

"Nothing's confirmed yet," said the first cop. "Don't jump to any conclusions. It could've been a medical condition. These things happen. The coroner will do some tests and we'll certainly tell them what you've told us."

Mirabelle came in through the door, flushed from the cold.

"What is going on?" she exclaimed. "Emma? What's all this?"

Mirabelle sauntered over, clutching her pregnant belly. She was also a redhead and looked very much like me, except that she was a lot taller, with brown eyes instead of green.

I hugged her and explained everything.

"Wow," she said. "Are you okay?"

"Fine," I said. In truth, I was a little shaken up.

"No you're not," Mirabelle said, looking deep into my eyes.

Tears welled up and I nodded. Mirabelle always saw right through me.

"It might've been my fault," I blurted out. "It was my drink. There was something in that drink that made her ill, I'm sure of it. If it's not the dairy, then—"

"Then what?" asked Mirabelle.

"Poison," I said. "Somebody was trying to poison me."

CHAPTER SIX

"Poison?" said Mirabelle. "No, Emma, come on."

"I know I sound paranoid, but crazier things have happened."

"Who would poison your drink in a busy cafe?"

"I don't know." I sighed.

Outside, the crowd was breaking apart as the ambulance drove away. The truck was in no hurry. She was dead, after all.

"A lot of people hate me," I said. "I've received more than a few death threats in the past."

"Oh, Emma, crazy people are always threatening celebrities. I'm surprised that you haven't had to file a restraining order on a stalker yet. But this is small town Ontario."

She was right. It was crazy for someone to do this in a busy cafe. If someone wanted me dead, wouldn't there be a better way for them to do it? But I still couldn't shake the idea away.

The cafe door opened and in stepped the ghost of misery's past. A man in a black wool coat with the collars turned up walked my way. I hadn't seen him in nine years, but he looked the same. The same moody gray eyes, dark hair, and broad shoulders.

Sterling Matthews. My heart sped up and I turned red.

From age fourteen to seventeen, we were practically inseparable. We met each other in the first year of high school, since we shared many of the same classes. At first we were awkward toward each other, but once we started getting to know each other on various class assignments, we became fast friends. In groups, he never talked much, but when we were alone, that was when he shared the most.

I used to feel privileged that he would share everything with me. About his family, about how his father had left the family when he was five. How his mother was working her tail off as a single mother of five. Of how exhausted he was to be the man of the house, taking care of his two brothers and two sisters when he wasn't in school.

When he had the time to sneak away with me when we were older, I'd try to devour him with my eyes and with my passionate kisses. He was my life. My parents adored him and his mother adored me. We were so young, but if he had asked me to spend the rest of my life with him, I wouldn't have hesitated.

The only problem was that he didn't feel the same. A month before we were to graduate, he started acting more distant. When he saw me, gone were the excited smiles. I could read his body language, and it seemed as if he was always turning away from me.

At first I tried to ignore it, writing it off as the stress of final exams and graduation, but July came. He worked in a grocery store, and I was home most summer days, recording music on my computer. We were both waiting for fall to go to a university that was only half an hour's drive away.

I wasn't thrilled about going to university, but Sterling planned to study Criminal Justice and I took English because the school didn't have a music program. He'd tried to talk me out of going to his school before. I'd thought it was because he wanted me to study something I cared about, but it turned out that he just didn't want me hanging around altogether.

At the end of July, I couldn't take it anymore. I asked him to meet me down by the lake. I'd shown up first and watched the ducks float along the waters. He came and hugged me from behind, yet I sensed that something was wrong. He had hugged me as if he was hugging me for the last time.

"Can I ask you something?" I asked

"Sure."

"Do you still want to be with me?"

I watched his face carefully. A dark cloud passed over his eyes and they were impenetrable. His silence told me everything.

It was then that I knew that we were over.

Then he tried to explain.

"I don't think you'd enjoy being at the same school," he said.

"Bullshit. You just want to date other girls."

Then he gulped. After a moment of silence, he took a deep breath and admitted it.

"You're right. We're too young to be tied down. It's better if we let ourselves experience different things."

"Or different girls, you mean," I said, standing up.

Sterling didn't come after me. He pulled out a fistful of grass and played with the blades in his palms, letting the grass and the soil stain his hands.

"I think you should do what you really want to do, not what I want to do."

He looked up at me and quickly turned away.

He had grown cold again, his shield against the world. I never thought he'd need it with me. I'd always been exposed to his soft side. But I'd been kicked out, and knowing Sterling, there was nothing I could do about it. He didn't want me and he wouldn't fight for me.

And I'd be damned if I stayed behind to waste my life in a small town and with a guy who didn't even care about me. Even though I was crushed, I did realize how stupid I had been. I was all set on throwing my dreams away so that I could be with this guy. I didn't want to study English. I didn't want to go to university. I had just wanted to stay by his side. As difficult as it was to lose him, I threw myself into my other passion: music. So I moved to New York and kept busy.

Friends kept telling me that I'd get over him as soon as I met someone else, but I always compared every guy I dated to Sterling. When he used to look

at me, it was as if I was the only person who existed
to him. When he hugged me, I felt completely warm,
secure and cared for. It was hard being sent back out
into the cold after that.

Then I met Nick. That was when I started forget-
ting Sterling more and more.

Now Sterling stood in front of me. We stared at each
other at a standstill, each waiting for the other to
strike first. All those feelings of anguish and obsession came flooding back. The feelings had always
been there, dormant; they'd never gone away. Last I
heard, he had gotten married and had two kids. That
was what stung most of all. He had long since
moved on.

"Hi," he said.

"Hi," I said.

"Mirabelle." He nodded to her.

Mirabelle smiled, but stepped back. Even she
could sense the heat in this room.

"What are you doing here?" I asked.

"I'm a detective now." He stepped in closer, under
the light. Yes. He was as gorgeous as ever. A dark five
o'clock shadow gave his face a more rugged look.

When he grimaced, two dimples appeared at the sides of his mouth.

"Oh. You'd always wanted to be a detective. I guess you got what you wanted in the end."

My voice came out sharp, even though I hadn't intended it to be so. But I couldn't put my pageant act on for Sterling. He'd know that it was fake.

"So how are you doing?" he asked.

"Fine. Everything's great. Couldn't be happier."

"Having a good Christmas with your family then?"

"Yes. It's fun. Except for this grisly death."

Sterling looked at the spot where the body had been.

"Yes." He frowned. "Not a pleasant homecoming, is it?"

"I've had better."

"I'd like to ask you some questions."

I crossed my arms. "I've said everything I needed to say to your colleagues, but now that you are here, I have something new to share."

"What is it?"

I told him about my poison theory as Mirabelle sighed in exasperation beside me.

"How likely is it?" I asked.

"We'll just have to see," Sterling said. He didn't say anything else. He didn't say whether it made sense or if he thought I was crazy.

His vagueness made me angry all of a sudden.

"Then this conversation is over," I said. "Mirabelle, I'll see you at home."

Without giving Sterling another glance, I stormed out.

CHAPTER SEVEN

*A*fter nine years, seeing Sterling still gave me anxiety. That guy still had a hold on me. The whole day had been one dramatic incident after another, and I was spent.

When I reached my street, I was crying again. How embarrassing to run into Kendra just as I was trying to wipe the tears away. She was walking her son home.

"Emma." Her eyes widened in surprise at the sight of me. "What are you—?"

"I'm sorry," I said.

Crying in front of other people was the worst. Not only was it incredibly humiliating, I hated to bring other people down.

"Is everything okay?"

I couldn't even look at her.

"It's nothing. I mean, it was just a crazy afternoon."

"Mommy, why is the lady crying?" her son asked.

"This is Blake Junior."

"Hi, Blake." I blew into a tissue.

"What happened?" asked Kendra.

I started sobbing again. My face was probably contorted into my ugly crying face.

"Maybe I shouldn't say this in front of your son."

"Say what?" she exclaimed.

When Blake Junior seemed distracted by a ladybug on the lawn, in a shaky whisper, I parlayed everything that had taken place earlier.

Kendra gasped. "Who died?"

"I don't know. This woman from another town. Her name was Emma too."

I heard Kendra inhale as I wiped my face with a new tissue from my purse.

"I think I need to leave," I said. "Coming back home was a bad idea. I've been here for less than twenty-four hours and somebody drops dead? It's not a good sign."

"Don't be silly," Kendra said. "I don't know what happened at the cafe, but you have your family here. And it's Christmas."

I looked up at her in gratitude for her kind words. It was unexpected of Kendra. Her face was still quite stony, but judging by the curves of her mouth, she looked like she was trying to cheer me up by forming what seemed like a smile.

"I read that your boyfriend broke up with you," she continued. "Is that why you're back?"

I nodded.

"That must be humiliating that he would dump you for some lingerie bimbo."

I cried even harder.

"You can't go back to that," Kendra said. "It's good to take a break from all that, don't you think? There are no tabloids here."

"Yeah, but there might be soon," I said. "As soon as word spreads that someone's been poisoned, it might be a media field day."

I didn't know why I was blabbering so much, but I couldn't help it. I was a wreck.

"Poison?" Kendra looked shocked.

"Mommy, what's poison?" asked little Blake.

"I'm sorry," I said. "There's no proof of that, so I didn't mean to say that in front of your son. It's just been a very crazy day."

Kendra nodded. "Right. It sounds stressful. Maybe you should go home and relax."

"Okay." I sniffed and wiped my last tear away. "Thanks Kendra."

With a heavy heart, I climbed up the porch stairs.

My parents had probably gone back to work after welcoming me. Alone in the house, I decided to hole up in my room and try to take a nap.

With the horrible things that had been happening to me lately, I thought it would be best if I disappeared from everyone. No ex-boyfriends, no tabloids, no fans, no dead bodies. But where could I go? Did I even want to go? Christmas was a time for family, and I'd probably be more miserable stuck on a resort alone somewhere. I could call a few friends who were living in Toronto or New York, but I didn't want to spoil their holiday cheer in my emotional state. Then there were my industry friends. They partied like crazy, but I didn't feel up for the party scene either.

I let myself drift off.

I dreamt that I was standing in the field at my high school and Nick and Sterling were both facing me. Suddenly they started throwing snowballs right at me. They were trying to kill me, telling from the menacing looks on their faces. I didn't know who had thrown the last one, but it hit my throat, and I began to cough out blood. The red blood tainted the pure white snow.

I woke up coughing and grasping my throat.

Maybe that was how the other Emma felt before she dropped dead.

My cell phone rang. My head felt groggy, but I

willed myself out of bed and reached into my bag. It was my manager, Rod.

"Hello?"

"Are you sick?" he asked. "I like that voice raspy. It sounds a lot sexier."

Rod was the kind of ostentatious New Yorker who wore fur coats and fedoras. He used to be a rock star, but retired to being a manager when the lifestyle got too fast for him.

"No, just been sleeping. What's up, Rod?"

"I'm hurt," he said. "I called Nick and he said you got out of New York and you didn't even tell me. I had to hear it from the tabloids that you guys had broken up."

"Yeah, well, it's not personal. It's not as if I went around telling everyone."

"How dare he cheat on you? I thought he was the good kind of bastard."

"I don't think he cheated," I said. "I broke up with him, but I really don't want to talk about that right now."

"Understood, honey. I just hope you're doing all right. Where are you now, anyway?"

I told him I was at my parents' house, but left out the part about the dead fan.

"Do I have a gig for you," he said. "A Middle

Eastern prince wants to pay you half a million to sing at his birthday party next week."

"What? You're joking."

"No joke. Completely serious."

"Who is he?"

"He wants to remain unnamed right now, but I talked to a representative who says he's a spoiled party boy who just wants to have a good time. His birthday party is being held in this crazy lavish hotel in Dubai. Don't worry, all you have to do is sing. No putting out or anything. I made sure of that. It's a five-song set on the twenty-fourth. It's in and out. Just a very long flight, but you get to stay in this gorgeous hotel and everything."

"Next week? But it's Christmas."

"I know, but it's half a mil!"

This sounded perfect. I was thinking about leaving earlier, wasn't I? But now that the opportunity was presented to me, I wasn't so sure.

"It's just that it's...Christmas."

"This is really a no-brainer. That is, if you're Jewish, like me."

I laughed.

"I'll think about it. Can I call you back tomorrow?"

"Sure thing, doll. Sleep on it."

CHAPTER EIGHT

*R*od was probably right. Who would turn down an offer for 500K to sing a few songs? I used to go around and beg for a chance to sing for people. But fortunately I was in a position now to make decisions where money wasn't a factor. And I also felt like I had a responsibility to stay.

The next morning I visited Mirabelle at her house. She was pregnant and she didn't need any more stress around this time of year. The cafe was closed until further notice and she was sitting in front of the TV, watching a cheesy soap opera and eating Cheetos.

"Want some?" She offered me the bowl.

"No, thanks."

I was apologetic about the cafe.

"Oh, what's to be sorry about?" she said. "I get a few days off. It's fine. These things happen."

"Do they?"

"Okay. It's not every day that someone dies in my cafe, but it's nobody's fault. I'm sure we'll find out soon that she just had some sort of medical condition. I think you worry way too much. You definitely got that from Mom."

"Why can't I be more cool and collected like you and Dad?" I sighed. "When do you think the cafe's going to be open again?"

She shrugged. "Maybe in the next few days. I should hear from the police soon. Once the coroner confirms that it was a health issue, we'll get the go-ahead to reopen."

I was still concerned that the cafe's reputation might be tainted by this, but I didn't say so. Mirabelle was so certain that there was no foul play, but I thought she was a little naive. She'd lived in this quaint town all of her life; she didn't realize that horrible things happened all the time. On my travels, I'd seen some crazy stuff.

"Ouch!" Mirabelle held her stomach. "The baby's kicking."

"Little Drew?" I called to her belly.

"Oh, the little rascal. Well anyway, just don't worry. Everything's going to be fine."

We watched the dramatic soap opera in silence. An evil twin was going to pull the plug on the good twin, who was in a coma, but she kept getting interrupted by visitors coming in and talking to the coma girl in the hospital room and the evil twin had to hide in the closet each time. The whole thing made me chuckle.

I didn't tell Mirabelle about singing at the prince's birthday thing. First of all, she would probably make fun of me. She might even encourage me to go, but I decided that I wanted to see her cafe reopened before I could say yes.

When I walked back to my family's house, a black Honda pulled up my driveway. Sterling got out. At the sight of him, I froze again.

"Hi," he said.

His two dimples made a brief appearance.

"Hello," I replied coldly.

"How you doing?"

"Fine."

He stood before me, looking as uncomfortable as I felt and trying to hide it.

"I'm sorry to intrude," he said. "I just thought it was best to tell you in person."

We were both hugging ourselves, freezing in the winter cold.

"Do you want to come in?" I asked.

"Sure, thanks."

My parents were both at work. Sterling looked around.

"Wow," he said softly. "It looks the same, like I'm traveling back in time."

"Wait till you see my bedroom," I said, but quickly regretted it. "I mean, do you want some coffee?"

"That would be nice."

I went into the kitchen to make a brew. It was really strange to have Sterling here. The bedroom comment had slipped out because we used to spend hours there, listening to music, hanging out and chatting. He'd been my best friend. I hoped he knew what I meant and didn't think I was insinuating hooking up. Which I didn't want. Certainly not.

I came out with two cups. One cream and two sugars. That was how we both liked our coffees.

He took a sip, then gulped the coffee down despite how hot it was. I stifled a grin, knowing that he liked it.

We sat on the couch next to each other, but I tried to stay as far as I could. Still, I could smell his cologne and I felt numbed by all the heat emanating from my nervous body. Whether he felt it too, I didn't know. He sat upright with his hands on his lap and looked everywhere but at me.

"I heard that you have kids now?" I asked.

"Yes, Maria is five and Sandy is two."

"Wow," I said. "I bet they're cute."

"They are." He nodded. "And I'm divorced."

"Oh...that's too bad."

Why he brought that up, I wasn't so sure. Mirabelle had already told me, so it wasn't a total shock. I wasn't sure how I felt about it. It certainly did hurt to hear that he had been married, that he was so in love with someone else that he took the plunge. But did I want him to be single? I wasn't so sure.

"So, did you find out anything new?" I asked.

"Well, yes," he said grimly. "I hate to tell you this, but the coroner came back with the results today and it's confirmed."

I held my breath. "What's confirmed?"

"She was poisoned. There was a lethal dose of cyanide in that hot chocolate. So you were right. Emma Chobsky was poisoned."

CHAPTER NINE

I exhaled. "I knew it."

"My partner has been researching Emma Chobsky's background to see if she had any enemies, but she doesn't as far as we can tell. She was working as a personal trainer, she was single, had plenty of friends, and was generally well liked."

"That would explain the soy latte," I said.

"What do you mean?"

"She drank soy because it was the healthier option. As you know, we took each other's drinks by accident because we both shared the same name but didn't know it at the time. So really, I was the target."

"Yes, which is another reason why I'm here. You're right. You're a celebrity. You're much more of a target."

"I get people saying nasty things about me and to

me all the time, but I didn't think that anyone would actually try something like this. I just don't know why."

"That's what we're trying to find out, but your safety is a concern right now. We have two guards monitoring your home at all times. They're in unmarked cars, but they are there in case you need anything."

"What, you think the killer is actually going to break in one day?"

"Who knows? But I do know that this house doesn't have a security system. You need all the protection you can get."

I shook my head. "I just never thought that something like this would happen in Hartfield."

"Horrible things happen everywhere," said Sterling.

"I feel terrible for putting my family and everyone in town in danger. What's going to happen to my sister's cafe?"

"It needs to remain closed for now, but our main concern is to get this guy behind bars as soon as possible. Our main concern is your safety."

He looked into my eyes. The way he looked at me was so soft and full of concern that it pained me.

I nodded. "So what's on the agenda to catch this killer?"

"Well, we've questioned everyone who was at the cafe. Unfortunately, there were so many people coming in and out that it was hard to keep track. The cafe has no security cameras, as you know, and there are none installed on the street."

"Any leads at all?" I asked.

"We have a list of the customers and we've questioned some of them and all the staff."

I could tell from the glint in his eyes that he had someone.

"Who do you suspect?" I asked.

"Well, what do you know about Cal?"

"The cashier?" I thought about it. "Nothing much."

"I questioned him earlier, and he set off some red flags."

"How so?"

"I asked him whether he saw anyone or anything suspicious. He must've thought I was insinuating that he did it."

I raised an eyebrow. "Did you?"

He shrugged coolly. "Whether I did or not, his reaction was suspect. He got defensive, saying that he would never poison you. But we're digging more into his life. He didn't grow up here. He didn't know anyone in Hartfield before he moved here a year ago. His family's from Toronto. Grew up there. Seems a bit odd that he'd pack up to move to a small town alone, don't you think?"

"It does, but what does he want with me? Kidnapping me for money or something, that makes sense, but trying to kill me?"

"Maybe he has some agenda with you. I want to look more into his background, see what I can dig up, but does he look familiar to you in any way?"

I thought about it. I met all kinds of people all the time.

"I don't know," I said. "All I know is that Mirabelle hired him a few months ago and she seemed to like him enough. I didn't meet him until the incident at the cafe. The line was so long that we didn't exactly have time to chat."

"Well, be careful. He's coming into the station and I'm going to grill him further. We'll get to the bottom of him. In the meantime, be on the lookout for anything suspicious."

"Okay."

"Here's my number." He slipped me his card with his phone number on it. "Can I take down your cell phone number in case I need to reach you?"

"Sure." I tried not to blush as I wrote down my number on a notepad from the side table.

Sterling stood up. I did, too.

"Thanks for your time, Emma. We'll be in touch."

locals didn't deserve to have their lives in jeopardy because of me. I had to help close this case as soon as possible.

I turned a corner and headed to the residential Swann Street, where Cal's place was.

A man was walking my way in the distance. As I got closer, I realized it was the man of the hour. Cal was walking towards me with his headphones on. He wore a grey ski jacket and had his hands in his pockets, looking melancholic. There was something tragic about him. His overgrown hair shaded one eye and he didn't seem to be aware of his surroundings.

My original plan was to break into his home, but now that he was before me, I could talk to him to suss him out.

"Cal!" I said loud enough for him to hear with his headphones on.

He blinked twice, and took off his earphones, fumbling to turn off his iPod.

"Hey, it's Emma," I said. "Mirabelle's sister?"

"I know who you are," he muttered.

"What are you listening to?" I asked with a smile.

"Oh, uh," he stuttered, "just music."

His eyes darted. He looked like he wanted to be here but in front of me.

I chuckled. "Well, fancy meeting you here."

"Yes. Good luck."

I could tell he wasn't sure whether to hug me and I just stood limp. He gave a small smile and left.

As long as we were just working together, I could stand it. Finding this crazy maniac who was trying to poison me took precedence over the awkwardness of reconnecting to an ex-boyfriend. Yet I felt woozy after Sterling's visit. This was the Sterling I knew and loved: strong, capable, caring. I tried not to let my body turn into jelly after I closed the front door.

I tried to concentrate on the case instead. Surely there was something that I could do.

I knew Sterling would do his best to get to the bottom of this because he always had a brilliant mind, but I had impatience on my side. If Cal wanted to kill me, there was surely proof of that, and I might as well go find it while the trail was hot.

CHAPTER TEN

I could see why Sterling wanted to a be a detective. If I weren't a singer, it would be fun to be a spy. It would be such an adrenaline rush. When I was little I used to spy on neighbours and eavesdrop on conversations all the time. And I was never caught. I figured that since I wasn't doing much, I could speed up the investigative process by helping.

I called Mirabelle.

"Hey, where does your employee Cal live?"

"Why?" she asked.

"I'll tell you later. I just wanted to have a chat with him."

"Are you up to something?" Mirabelle asked.

"No, I just want to ask him a couple of questions about the whole cafe incident."

"Fine, I'll forward his phone and address to you on the phone."

"Thanks. Hey, does he live alone?"

"Yes, I think so. Why you need to know?"

"Just wondering. Gotta go." I hung up.

There was no need to sit and wait for answers when I could just go to the source. While Sterling had to play by the books, I could get down to business and do the real digging.

When Nick was training for his action film, went to visit him in Queens for his training sessio I even trained with his trainer for fun and now I very adept at Krav Maga, this brutal form of cc and self-defense used by the Israeli army. So I afraid to be attacked; I could handle mysel was the killer, I could take him. I might lc but I could pack a punch.

Cal lived only three blocks from the third floor of a little apartment buildin clothes on—all black for the mission into a bun, and headed out.

I passed the Chocoholic Cafe but the sight of the place closed of the bustling street made me sa whoever did this. I had to. A merry shoppers, the charm stores, the lovely residents –

"I live here," he said, after a beat.

"I was just on my way to see a friend. Where are you off to?"

He buried his hands deep into the jacket pockets and looked down at his shoes.

"The police station," he said. "They have some questions about the, you know, problem at the cafe."

"Why would they ask you?"

He shrugged. "I guess they think I might know something, since I talk to everyone who comes in."

"They don't think you had anything to do with the death do they?" I asked casually, like a joke.

"I don't think so," he said seriously. "Because I had absolutely nothing to do with it."

Did I detect a note of defensiveness in his voice?

"Well," I said. "I'm sorry that the cafe's closed down. It's such a shame for everyone."

Cal nodded. "Yes, but I'm not worried about my job. You know, from what I gathered from the detective, someone might be out to get you. Are you afraid?"

He blinked twice and stared at me as he waited for my response. I felt uneasy, but I stared back, as if to answer a challenge.

"No," I said. "I have nothing to be afraid of. The only thing I'm wondering is why?"

He shrugged again. "I don't know, but there's a lot of psychos out there, you know?"

"Yes," I said.

His face was still blank and inscrutable. He turned to leave.

I could see why Sterling had his suspicions. Cal was a bit off. He was definitely secretive. I felt as if there was something he wasn't telling me, which was why I planned on breaking into his apartment.

I once read that you could tell more about someone by looking around his room than talking to the person. I really believed that. I mean, if you went through my room, you'd see how girly and romantic I was. If this guy was a killer, I'd know it by a glance around his living space.

Snowflakes slowly descended from the sky. The street was pretty calm except for an old man who went into Cal's building with a bag of groceries. I knew that Cal would be at the police station for a while, so I had plenty of time for a quick check-in.

I could've followed the man in, but there was a much easier way. This was Hartfield, a place ridiculously easy to break into. If Cal was like everyone else in town and lived in a house, he probably would've left the front door unlocked, as many of my neighbours and my parents did.

I quickly went around the side of the building to the back. The building had a fire escape, and I had to jump to pull the stairs down. I climbed as quietly as I could so that the neighbours on the other two floors wouldn't be disturbed, if they were there at all.

On the third floor, I peered into the window. The curtains were open, and nobody seemed to be inside. Of course the window was unlocked. It was heavy, but I managed to lift it.

When the opening was big enough, I slipped in. I stood in Cal's living room.

It seemed like a place that a bachelor in his early twenties would live in. Mismatched furniture was carelessly arranged in the space, including a raggedy couch in an ugly shade of brown and a beat-up oak coffee table. A bong was on top. He owned a modest-sized TV and a collection of video games. There was an acoustic guitar by the corner. The fridge contained more condiments than real food.

I found a notebook on the kitchen counter. I was hoping it was his journal. I flipped through it and read a few pages. It was all poetry. Really cheesy poetry, the emo kind I used to write in high school. Many were about heartbreak and love. Poor kid.

Looking around, it didn't seem as if he could be the killer. Maybe we were wrong. Maybe he was just a sensitive kid who wanted to get away from the city for a while. Find himself and all that. I could relate to that. In fact, looking around the sorry excuse for a living room, I felt sorry for him.

Next I checked out his bedroom.

And that was when I freaked out.

CHAPTER ELEVEN

*M*y own face stared back at me from every angle of Cal's bedroom. Posters, magazine shoots, album covers, tabloid pictures, you name it. All my CDs were stacked on top of his stereo. He even owned all my singles and EPs. He was also the proud owner of most of the merchandise we sold from my last North American tour: T-shirts, buttons, tote bags. It was totally creepy. Flattering, but mostly creepy.

I had never seen anything like this. Sure I'd met many people who had claimed to be my biggest fan, but I doubted that they slept in a shrine devoted entirely to me.

What I didn't understand about obsessive fans was why they would turn on the very person they were obsessed with.

But what did I know? Sometimes the line between love and hate was a bit blurred.

I called Sterling.

"Emma?" he answered.

"Is Cal still at the police station?"

"Yes," he said. "I'm in the middle of questioning him, why?"

"Don't release him. I'm in his apartment."

"Why are you in his apartment?"

"I broke in," I said casually.

"Emma, are you out of your mind? You can't just break into people's houses!"

"Well, he wasn't here, was he? Do you want to know what I found or what?"

I told him about the Emma Wild shrine that I found. However there was nothing further to go on. I searched his drawers, bed, closet, everywhere. No poison, no weapons. The guy didn't even own a computer.

"Get out of there," said Sterling.

"Just keep him there a bit longer," I said. "I need to find something."

I hung up and kept searching. Maybe Cal had a secret hiding spot, like under a plank of wood on the floor or something.

After ten more minutes of searching, a voice

piped up from the window. I jumped, startled by the deep voice.

"Emma." It was Sterling, looking a bit frazzled. "You nut. Do you have any idea how dangerous this is?"

"Oh, hi, Sterling." I smiled sweetly.

I was starting to get used to his presence. Sure, he had shattered my heart once, but maybe over the years I'd overblown that incident to monumental proportions. I'd always been a dramatic person. I was even comfortable enough to tease him a little.

"Are you my father? Relax, it's fine. I know self-defense."

He sighed and rubbed his face. "Let's see it then."

I showed him the room. He was impressed.

"A bit excessive, huh?"

"It's looking a lot clearer now," he said. "This might just be our creep."

"But the thing is, we don't have much proof. So far, he's just an obsessed fan."

Sterling crossed his arms. "There might be more to it. Apparently he used to have a drug problem when he was a teen. He even went to juvie for a few months after fighting with someone and almost pounding the other kid to death. He's obviously a troubled guy, possibly mentally disturbed."

"Why would he come after me?"

"For control. Maybe he's obsessed with you and knows that he can't have you. After all, you are with this hotshot movie star. Maybe he doesn't think he has a chance and would rather kill you than lose you to someone else."

He said it so matter-of-factly that I almost laughed.

"Seriously? That's what you can come up with for a motive?"

Sterling shrugged. "Just a guess."

This time, I really laughed.

"I just can't imagine it."

"Imagine what?" he asked.

"That someone would be so in love with me to want to kill me."

He pressed his lips together, staying silent for a couple of minutes, then he turned to me.

"Emma, maybe you don't know the effect that you have on people."

The humorous smile faded from my face. Sterling looked serious. And sad. Could he still have feelings for me?

"You're beautiful and talented. Your music is enormously popular. Certain guys will put you on a pedestal. It's not a good feeling when they're faced with something that they can't have. When they're so close."

I slowly nodded.

"You know," I said. "My life's not that perfect. People think I have this lavish life, but I'm still the same person. There's just more people looking up to me. And I kind of resent that sometimes. I can't be this perfect role model all the time. Because frankly, they don't know how much of a mess I can be. How much of a mess I am. The media can adore me and easily turn on me, and that's the case with this case. Love can easily turn on itself, can't it?"

Sterling sighed.

He took me into his arms and hugged me.

He didn't need to say anything. These were the kind of moments when I really enjoyed his silences. Because I knew that he understood. And if he didn't, he was willing to try.

I pulled away.

"And my perfect relationship to the famous movie star? Well, that's over. Gone in a New York second, okay?"

"I'm sorry," he said softly. "What happened?"

"He wanted lingerie models."

The somber way I said it made him chuckle. "He's a shallow idiot."

"I agree."

"It's agreed then," Sterling teased. "Your life is terrible."

I scowled. "Yeah, yeah, I know. I hate it too when a celebrity complains. I know how good I have it."

"Someone's trying to kill you," Sterling said. "You have the right to feel a bit apprehensive about your lifestyle."

"You're right. Thanks." I headed back out the window. "You coming? I really want to talk to this guy and get to the bottom of it already."

CHAPTER TWELVE

"So I hear that you're a big fan of Emma Wild," said Sterling.

I watched Cal in the interrogation room from behind the one-way mirror as Sterling questioned him. Cal's eyes widened the way they did when he'd run into me earlier, as if he'd been caught doing something he shouldn't.

Cal didn't answer.

"Are you or are you not a fan of Emma Wild's music?"

He nodded. "Yes, I am."

"Do you have all her CDs?"

Cal paused. "Yes. I listen to quite a lot of her songs. That's not a crime, is it?"

"You just don't seem like the type."

"The type to what?"

"Listen to her music."

Cal looked offended. "And what's wrong with her music? It's well written, her voice is beautiful, and it's deep. She gets me."

Sterling crossed his arms and narrowed his eyes at him.

"Seems a bit suspicious that you'd come all the way to live in her hometown, where you know no one, and get a job at her sister's cafe. Is that what you came out here for? For a chance to meet Emma Wild?"

"No. I didn't think that would happen. I thought she lived in New York."

"But you knew that her family lived here, right?"

"Yes, but—"

"And what a coincidence that you got a job at Emma's sister's cafe. What a better chance to meet Emma than there, short of breaking into her home, right?"

"Really, I had no idea that Mirabelle was her sister until I got the job."

Cal's face turned red. Was he embarrassed or guilty?

"I want a lawyer," he said.

"You have plenty of experience with lawyers, don't you?" Sterling continued. "You had a slew of trouble when you were in your teens, and now you're on the straight and narrow?

"Yes," Cal said through clenched teeth.

"It just so happened that you abandoned the city where your friends and family live to move all the way to a small town where you know no one."

"Yes. If you really must know, the city was getting too crazy for me. And my family and I don't get along. It's true that I'm a big fan of Emma Wild. I knew she'd grown up here, and it sounded like a nice town, so I moved up here, to get away from everything. And it really was a coincidence that I got a job at her sister's cafe. Except that I wouldn't call it a coincidence."

Sterling leaned in on the table. "What would you call it then?"

A dreamy smile spread over Cal's face. "Fate."

Sterling raised an eyebrow.

"Fate wanted us to meet." Cal's face lit up the more he spoke. "And when we finally did, I was completely starstruck. I mean, I've been to so many of her concerts and I listen to her music all the time. In fact, I was listening to her last album when I ran into her on the street earlier. Imagine that! Emma Wild running into me and wanting to talk to me. I was so embarrassed that I didn't know what to say to her. I was afraid that whatever I said would make me sound like a blubbering fool so I said next to nothing."

"Why do you love Emma so much?" asked Sterling.

"My upbringing wasn't so good. It was true that I got into all sorts of trouble as a kid, that by the time I was twenty, I couldn't take it anymore. If I'd stayed in the city, I would've gone deeper with the wrong crowd. My drug addiction would've gotten worse. I needed to clear my head. The last straw was when my girlfriend broke my heart, man. I was in love with her for three years.

"We were finally together when I got my act together, but then she dumped me. Dumped me for some crappy asshole who works in I.T. Emma's music was what got me through it all. She saved my life. If I didn't have her music to get me through the horrible times I went through in the last few years of my life, I wouldn't be here right now. I travelled, worked on the road, went to her concerts, got my head straight. And now I'm here, in Hartfield. And I love it here. It's so peaceful. The people are so friendly."

Sterling seemed to be digesting his story.

"Look," said Cal. "I would never hurt Emma, I swear to you. I'm her biggest fan. Why would I hurt her?"

Sterling sighed. "Somebody I spoke to thought you were involved, or might know something."

"Who?" Cal frowned.

"I'm not at liberty to say, but you do come in contact with everyone at the cafe. You do touch every cup when you write the customer's name on them."

"Yes," Cal said, exasperated. "I do write the names on the cups, but the baristas are the ones who make the drinks. They would know if the cups were filled. I don't leave the cash register once I give the cups to them. In fact the counter is quite far from the cash register."

"Do you think one of the other workers had something to do with it?"

"No, I don't think so." But something in Cal's expression told otherwise.

"Are you sure?" Sterling pressed.

"Except..."

"What?"

"Kate, one of the baristas. She hates Emma's music."

"Why does she hate Emma's music?" Sterling asked.

Cal thought about it. "I think she mentioned that she just didn't like her voice. That it gave her a headache. Oh, and she was the first one to leave after we were all questioned. She looked really ill and really shaken up. It could've been that she was shocked, but who knows? Maybe she was overdoing it as an act."

"Well, what do you think?" Sterling asked me.

"I believe Cal. He didn't do it," I said. "What do you think?"

"I don't think he did it either."

Cal was just a hardcore fan who was going through a difficult time. I was glad that my music was able to help him. Plenty of fans wrote to me to tell me about their breakup stories all the time. When you were hurt and vulnerable, you needed someone who understood, and I was the safe person who was close but far, who whispered sweet songs of comprehension in their ears and gave them reassurance that they weren't alone.

"I'll go visit Kate at her home," said Sterling.

"I want to come," I said.

"No. It's better if I do it. Don't want to put you in danger."

I sighed.

"Fine," I said. "I'll go home and ask Mirabelle what she knows about her."

I didn't know Kate well, but the few times that I'd

talked to her, she seemed nice and friendly, but so were those journalists who used to interview me. To my face, they were adoring angels. Then later I'd find out that they'd written the most scathing review or the most damning interview. In all my years in the industry, I knew that people weren't to be trusted at face value. And I didn't trust Kate.

CHAPTER THIRTEEN

*M*irabelle was on the couch again, finishing the stale cakes from the shop ever since it closed down. Somebody had to finish the inventory. On TV the evil twin on the cheesy soap was caught in the act of trying to kill the good twin by her mother. The mother was in hysterics.

"So what's up with Kate?" I asked. "What's the scoop on her?"

"Why? Is she a suspect now?"

Mirabelle had already been informed by the police that the drink had been tempered with. I didn't try to rub in the fact that I was right, seeing what a terrible situation it was to gloat about.

"Well, maybe. I mean, she does hate my music."

Mirabelle's jaws dropped. "Really? Says who?"

"It doesn't matter."

"I hope that's not true. I play your music in my cafe all the time."

"Thanks, sis."

"I hope it's not Kate. It can't be. She's so nice."

"What do you know about her?"

"Well, she's two years younger than you."

"Did she grow up here?" I asked.

"Yes. But you've probably never seen her because she had twins at sixteen and had to drop out of school."

"Really?" I exclaimed. "Twins? Wow."

"No kidding. She'd been working odd jobs since then. Never graduated high school. Her boyfriend, the father of the kid, skipped town five years ago. Terrible. So I was happy to give her a job at the cafe just when she was broke and getting kicked out of her apartment."

"Are you close at all?" I asked. "Does she talk a lot about her personal life?"

"Well, she talks about the kids, and sometimes who she's dating. So yes, I guess you could say that we're friends. She's closer to Kendra. She's her cousin, you know."

"Kendra? I didn't know they were related."

"I don't think they were close growing up. Kendra used to distance herself from Kate because she got

pregnant so young and Kendra didn't want to be associated with all of that. But now, as far as I can tell,
they're pretty close. Kendra comes in to chat all the
time."

"I need to go talk to Kendra then."

Mirabelle put down the piece of cake and gave me
a look.

"You're not going to tell her that Kate's a suspect
are you? I mean, you can't just accuse her cousin of
trying to poison you."

When I knocked on Kendra's door, I heard Blake
Junior screaming in the house.

Kendra was shouting at him to pipe down when
she opened the door.

"Hi," I said. "I hope it's not a bad time."

The kid was banging on a pot and singing at the
top of his lungs in the kitchen.

Kendra was surprised to see me. She hesitated at
the door.

"No," she said. "Come on in."

She cringed when her son sang a high note.

"Blake! I'll give you a dollar if you go upstairs and
keep quiet."

"A dollar? Yay! I can buy ice cream!"

The kid with chocolate on his face happily went up the stairs.

Kendra rubbed her temples. "Ugh. I have a headache."

"How old is Blake?"

"Four. It's hard to keep up with him, and he still has all this energy even after running around in daycare all day."

"I didn't mean to intrude," I said.

She narrowed her eyes at me. "Why are you here?"

"We've been trying to get to the bottom of this whole case at the cafe, as you know. And it has been confirmed. Someone did poison the drink."

Kendra looked shocked.

"Oh my God."

"Yes. Cal mentioned that Kate might know something and I was wondering if she told you something?"

"Kate? No. What would she know?"

"We just don't want to leave any stones unturned, that's all. Kate's at the police station, being questioned right now. I just wanted to know if she'd mentioned something strange. She's closest to all the action, and maybe she mentioned whether she suspected someone, or knew something that she wasn't telling us?"

"Or if she did it herself you mean?" Kendra asked. "Is that what you're trying to say?"

"No. Of course not." I chuckled.

"Why would she want to kill you?" Kendra had her hands on her hips. Oh lord.

"I'm not accusing her."

"Sure, she has a temper sometimes. Sure, Kate hates your music and your looks and the fact that you stole her man, Nick Doyle."

I gasped. "What?"

"She's obsessed with Nick Doyle. She loves all his movies."

"Oh."

"And she's jealous that you're with him. Plus, sometimes Kate can get a bit stressed out. Her life is in shambles, has been ever since she had those twins young. Losing the best years of her life to raise kids? Brutal. She's had more than a few nervous breakdowns. Boyfriends just come and go. The baby daddy's long gone. And she sees you and your perfect life, and well..."

"What, so she does hate me?"

Kendra paused. "Hate is a strong word."

I was expecting Kendra to defend her cousin, to say that she was sweet and caring, and that she would never hurt a fly. But this just got interesting.

CHAPTER FOURTEEN

I called Sterling, but he was probably still questioning Kate. After getting her address from Mirabelle, I set off for her house.

I had to pass through Samford Street because she lived on the other side of town, but a crowd of paparazzi had shown up. They were at the cafe, taking pictures.

My instinct was to run away, but I was so surprised that I froze and watched them.

Until it was too late. One of them saw me and began running towards me.

"Hey! Emma!"

"Is it true that someone was trying to poison you?"

"Who do you think is the killer?"

"Is it true that Nick dumped you for that hot model?"

They had cameras, and one of them even had a video camera. I began to run. For a small woman with short legs, I could run really fast. As I did, my phone rang.

"Emma," Sterling said. "What's going on? What's all that noise?"

He was referring to all the commotion behind me. The townspeople were looking at me, while the hound dogs were hot on my trail.

"Paparazzi scum," I said. "They're here."

I explained that I was literally on the run as I spoke.

"Where are you?" he asked.

We made plans to meet two blocks away, so I ran even faster. All the time at the gym working out paid off. I silently thanked my trainer. By the time I reached my destination, those scum were eating my dust.

Sterling pulled up in his black Honda. My getaway ride.

"Thanks." I jumped in and he sped away.

"How did they get here?" he asked.

"I should ask you. You're the detective."

"It could be anyone," he said.

"That's how it always happens. Where should we go?"

"Well..." Sterling raised an eyebrow. "Are you hungry?"

He drove outside of town until we were surrounded by farm land covered with snow.

"I know this great diner that just opened up," he said. "I bet you've never been."

After a few more minutes, he pulled into The Burger Shack.

"This is nice," I said. "Away from all the commotion."

We slid into a booth. When the waitress came, I ordered a cheeseburger and a chocolate milkshake, and Sterling ordered the same.

"I see your taste buds haven't changed," he said.

I made a face. "What, you think I would live off salads with low-fat dressing now?"

"Isn't that the celebrity diet?"

"Yeah, well, I do admit to that, although I reserve my weekends for junk food. I also have a dozen people working on me at all times. It's all part of the job unfortunately."

"You look great either way," he said. "But I always liked you better with no makeup."

"Thanks. I do too."

We shared a moment of looking into each other's eyes.

"So this is your life, huh? Is it always that chaotic?" he asked.

"Worse," I said. "There's usually one or more of these paparazzi bastards trailing me whenever I go outside. Which is why I stay in a lot of the times. I thought I'd get some peace in Hartfield, but they found me."

"Don't worry." He grinned. "I'll protect you."

I grinned back.

"Did you talk with Kate?" I asked.

"Yes. But I didn't get much out of her. She said the reason she left so abruptly that day was because she was sickened by the whole thing and went home because she was nauseous."

I told him what Kate's cousin told me.

"Unstable, huh?" he said.

"Yes. Do you think that she has access to cyanide?"

Sterling thought about it. "I thought I saw her some time ago with this guy who worked at the pharmacy. But they broke up a year ago."

"But she might still have access to it?"

"Maybe. Sure."

"But I still don't understand her motive. Does she hate my music so much?"

. . .

"Maybe it is jealousy," Sterling said. "She's a single mom of two kids with no education and tons of debt. She has to work two jobs. And here you are, wealthy, loved, dating rich movie stars."

His eyes clouded over. Maybe Sterling was the one who was jealous. If only he knew how much I was in love with him, how hurt I was when he dumped me...

"That's what Kendra said," I said.

The waitress brought our food and we dug into our burgers. We were both famished.

"I want to search Kate's house as well," I said.

"Let's not have another B&E," he said. "I'll get a warrant this time."

CHAPTER FIFTEEN

The next morning, Sterling and his team went over to Kate's house. I wanted to be there, but he wouldn't let me come.

It was a Saturday morning, and I stayed in baking gingerbread cookies with my parents instead.

By the time the cookies were in the oven, Sterling called me, excited.

"It's her," he said. "We just found the bottle of cyanide hidden inside her toilet."

"No way!"

"She's been arrested by the police."

"So you'll be at the station soon?" I asked.

"Yes, but—"

"I'm coming."

Sterling sighed. "I guess I can't stop you, can I?"

"Don't you know? I'm a celebrity. I can get in anywhere."

"I bet you can."

"Come on, please?"

"Fine. Come on over. Maybe you'll be an asset."

"It's not me! I swear," Kate shrieked as they brought her up the steps and through the front door of the police station.

Her brown hair was a mess over her face and her hands were handcuffed behind her back. She sobbed and snot dripped down her nose. Once she was inside, the police uncuffed her and Sterling gave her a tissue.

"Emma!" she said when she saw me coming in the front door. "I didn't do it. You have to believe me."

I signalled to Sterling to bring her into the interrogation room. I wanted to talk to her, too.

"If you didn't do it, why is there a bottle of cyanide in your house?" I asked.

"I was framed."

"By whom?"

Kate cried louder.

"Who, Kate?" My voice rose.

"Kendra."

"Kendra?" I gasped. "Why?"

"I wasn't sure if it was her at first, but nothing else explains it."

"Start at the beginning," Sterling demanded.

"I remember seeing Kendra that day at the cafe. She was just leaning in to chat, said something about cooking dinner the next day. It happened so fast that I didn't think of it."

Kate blew her nose.

"So you didn't see her poison the drink," I said.

"No, but when the woman died, I got suspicious. I went to Kendra's house immediately after to ask her about it."

"Why did you immediately jump to the conclusion that your own cousin did it?" Sterling asked.

"Yeah," I added. "I thought you were close."

"We are," said Kate. "Which was why I suspected her. You see, I know how much Kendra hates you, Emma."

That stung. "Why does she hate me so much?"

"She blames you for her husband's death."

"What?"

"She has to blame someone and you were easy. The thing is, her husband killed himself. He was depressed for a long time and he used to listen to your music all the time. He said you had a voice like velvet. Kendra hated that even when he was alive.

They used to get into fights when he listened to your music in front of her. She was so jealous. He said that you were the only one who understood him. Then finally, he poisoned himself and left a note, which Kendra destroyed."

"Wow." I leaned back in my chair. "What did the note say?"

"I don't know," Kate continued. "She only told me about it, but not what it said. She didn't want anyone to know that he had committed suicide, which was why she didn't even hold a funeral. Since that day, a bitter seed was planted in her heart and it grew bigger and more powerful every day. Every time I saw her she mentioned how much she hated you and wanted to kill you. I thought she was just venting. I really didn't think she was serious. I know that deep down she blamed herself, but it felt good for her to blame you."

"So when you talked to her, what did she say about it?" asked Sterling. "Did she admit to it at all?"

"Kendra was totally defensive about it. She played dumb and said she didn't know what I was talking about. But I knew her husband had poisoned himself with cyanide and she still had the bottle. I just didn't know that she would turn around and frame me. All I'd ever done was try to protect her."

Sterling shook his head.

"You wanted to protect her so much that you were willing to put Emma's life and the lives of other people in town in danger? You should've come forward with this information sooner."

Kate sobbed.

"I knew how much she'd been through. I was her maid of honor. I thought we were close. I can't believe she'd frame me like this. Why?"

Sterling and I looked at each other.

"We'll have to ask her, won't we?" I said.

CHAPTER SIXTEEN

When they brought Kendra in, I decided to bring her a cup of hot chocolate from the station's lunchroom.

She looked up, surprised to see me. I gave her a smile that she didn't return.

"What's the matter?" I said. "You don't like hot chocolate?"

Kendra stared at me with that chilling dead look of hers.

"Not your drink then." I took a sip of it. "But you know it's mine."

This was the first time I drank hot chocolate since Emma Chobsky was poisoned. I had to say, it was delicious, but of course it wasn't nearly as good as Mirabelle's.

"How long have we known each other?" I asked.

Kendra shrugged. She looked lifeless sitting in that chair. She acted as cool as a cucumber, but I knew she was scared.

"Since we were eight," I answered for her. "I remember the first time I met you. You had your own little clique, your band of followers even then. You almost tore my hair out for sitting at your lunch table."

"It was an exclusive table," she replied. "And I never gave you permission to sit there."

"Little Kendra. Always a control freak. When did Blake die?"

"It will be seven months this Saturday," she said. "Can I go home? I'm getting tired."

"Tired from what? You don't do anything all day. Blake's insurance gave you a good chunk of money."

"Well money doesn't buy happiness," she snarled.

"Why did he die?" I asked.

I could see the steam rising to her face. She turned red, seething.

"He was sick," she finally said.

"What was he so sick from?"

She composed herself. "I'm not going to talk about this."

"It was poison, wasn't it? Cyanide?"

Her face was beet red, but she still wasn't talking.

"You came home one day and your husband was

found dead by his own hand in the bedroom. My first album was playing when you got there. How did that make you feel?"

Kendra shot me a hateful look. "How would you feel if the love of your life died?"

"It would be excruciating. Is that why you tried to kill me? Did you hate me that much?"

"No. I didn't try to kill you."

I locked eyes with her.

"Eyes don't lie, Kendra. There's hate in your eyes and it's directed at me. Now isn't it funny how someone tried to poison my drink with cyanide? The cyanide that remained from your husband's suicide?"

"It wasn't. You already got your perp. Kate."

"Your own cousin? Really, Kendra?"

"Kate's a liar! She belongs in a psychiatric ward!"

"For what?" I spat out. "For trying to protect you? I agree. She is crazy for trying to cover up a murder in your family. Can you imagine how Emma Chobsky's family feels?"

"That was an accident!" she blurted it out.

It was too late for her to take that back. The jig was up.

"Come on, Kendra. You framed your own cousin. You tried to poison and kill me in public — how did you ever think you were going to get away with that?"

Kendra stood up. I was wearing my flats and she loomed over me.

"It was all your fault," she spewed out. "The moment you stepped into town I wanted to rip you apart. It wasn't planned. In fact, I would've tried to kill myself if I didn't have Blake Junior to take care of. I hated you so much for what you've done to my family. So I followed you. The cafe was so crowded that nobody saw a thing when I poured the poison, especially when people were distracted by your screaming admirer."

"Whom you killed," I retorted. "You killed an innocent young woman."

"Like I said, it was an accident!"

Kendra lunged toward me and Sterling burst in and stood between us.

"Sit down, Kendra," he said.

"My husband is going to hell because of you," Kendra continued. "Suicide! Nobody gets into heaven with suicide. And you did it. You made him do it with your depressing songs. He kept replaying that one song, 'Die For You,' over and over. And the other one, 'The Killer in Me.' I thought I was going to go mad, too. Are you proud of that? Are you proud of what you've done?"

The blood drained from my face. To be accused of being responsible for someone's death was so vicious. Kendra could see that it affected me, and that encouraged her.

"I was doing the world a favor. How many more people are going to kill themselves because of you? Shame. Shame!"

Sterling pushed her to the wall. Two policemen came in and cuffed her again.

Sterling saw how shaken up I was and he embraced me.

"I was doing God's work!" Kendra railed. "Blake and I loved each other. We were supposed to grow old together. My son is fatherless because of you. All thanks to you and your devil's music. Go to hell, you murderous witch! I still have another plan to kill you. To bury you. Go to hell!"

Sterling took me outside, but I could still hear her screaming in that room.

I sobbed in his arms.

"Don't listen to her," he said. "It's not your fault."

"It is partly my fault. My music triggered something in him that—"

"Shhh, Emma. Your music brings joy to so many people. Blake was mentally ill. I'm sure Kendra blames herself more. She's just taking it out on you. I'm sure she felt as if she failed as a wife. If they were so in love, why would he kill himself? There was nothing she could do. She was helpless. She just needed someone to blame."

I nodded. I knew that he was right, but I still couldn't shake off the guilt.

"What you have to offer the world is amazing," Sterling said. "You're amazing."

It did feel better to be in his arms. I felt protected. I'd always felt warm and safe in Sterling's arms. God, how I'd missed him.

CHAPTER SEVENTEEN

Kendra was locked up and awaiting trial. The paparazzi stuck around and were all over the scoop while I holed up at home and ordered Christmas presents online. It had been a crazy December. Emma Chobsky's funeral took place last weekend. I was invited and I decided to go out of respect to pay tribute. I even sang a song. The family wept. So did I. I was pretty glum for a few days, but things cheered up when we got closer to Christmas.

I tried to focus on the positive. I had a lot to be grateful for. The killer was behind bars, I was safe, and I had my lovely family to celebrate with.

On Christmas Eve, most of the paparazzi had left. Hopefully they had family of their own to spend time with.

Mirabelle and her husband Sam were over at my

parents' house. After our traditional Christmas meal where we stuffed our faces with all sorts of comfort food and desserts, we carried the party into the living room to drink eggnog and open presents.

"This one's for you, Mom." I passed my present to my mother and watched her open the pink box.

"A new knitting set! Is this real silver?"

"Yes."

My mom was in a knitting circle, and I figured this would score points with her ladies. Who didn't envy someone with the hottest knitting needles?

"You're next." Mirabelle gave me a red box.

I shook it first, then sniffed it.

"It doesn't smell like chocolate," I said with disappointment.

"It's better," she said.

I opened it. "The Super Kid's Detective Kit!"

It was a set for young sleuths in the making. It came with a spyglass, a notebook, finger print powder, and a book of codes.

"Thanks! I love it. I'll put it to good use. Didn't I have one of these as a child?"

"You did," said Mirabelle. "That's why I got you an updated version. You used to drive everyone crazy trying to dust fingerprints all over the furniture with all that white powder."

I laughed at the memory.

"I always wanted one of those Sherlock Holmes hats," I said. "And the cape to match."

"Well, that's a present for next year," said Dad. "Come on, let's sing some songs."

He began to play the piano rendition of Wham!'s "Last Christmas." It made us all cringe at first, but we sang along anyway. When Dad transitioned into Mariah Carey's "All I Want for Christmas is You," the doorbell rang.

I stood up to get it. When I opened the door, boy did I get a nice surprise.

Sterling.

"Hi," I said.

Those old butterflies came fluttering back into my belly.

"Hi." He grinned.

"I hope you're here to give me my present."

"I am."

"Come on in."

He pulled at my arms. "I can't stay long. I just came to see you."

"Oh?"

I stepped out and closed the door behind me.

It was snowing lightly and the street looked so peaceful and gorgeous.

Sterling looked very handsome in his dark coat. His bright grey eyes shone under the lights.

"So where is it?" I asked. "Where's my present?"

"It's here."

He pulled me in. His lips softly touched mine. One hand lightly touched the small of my back as the other rubbed below my neck. It was the kiss that used to make my head spin as a teenaged girl.

When he pulled away, my legs felt like jelly.

"Emma, I know that after Christmas you might be going back to New York, so I have to tell you."

I looked at him expectantly. "Tell me what?"

"I know that I've been an idiot. It's hard for me to express myself. I know that. It's the reason that I got divorced, well one of them anyway. It's also one of the reasons I lost you to begin with. I know I need to learn how to tell people how I feel, and now I'm telling you."

I gulped and waited. The smell of him was intoxicating. My heart pounded like crazy.

"You know why I broke up with you?" he asked.

I frowned, not wanting to think about it. "Because you didn't love me and wanted to date other girls."

"No. That wasn't the reason."

"So what was it?"

"I did love you," he said softly. "I was crazy about you. I knew that you could do whatever you set your mind to, and I knew that you'd stay in this town for me. Except that I also knew that you had more to offer the world. So I had to let you go."

"That's ridiculous. So that was why you were so cold? So I would think that you didn't love me?"

Sterling nodded. "And I was right to. You know? You had all that success—success that I knew was in you. You'd always been so passionate about music and you couldn't sing at birthday parties for the rest of your life. I'm glad I don't have that on my conscience —the guilt that I had held you back."

I took a breath. "Sterling, that is just so cruel. How could you? You broke my heart. All these years I thought you didn't love me. And I wrote all those songs about it."

"I'm sorry. I wanted the best for you and I just thought this town was too small for you. I wasn't enough for you."

"How could you say that?" I exclaimed.

"I thought that you would get over me and be with someone else, someone on your level. And you did."

"But I spent years trying to get over you. You don't know how much I cried and suffered over you!"

Now it was his turn to look surprised. "Really?"

"Yes. I wrote so many songs about our breakup."

"I didn't know that you felt so strongly. When you came back, I was unsure at first, but being around you again…I knew that I didn't want to let you go again.

I laughed. "I can't believe it. I mean, I really thought that you didn't care. I avoided this town because of you! You really are cruel. All those years of torture. You know that you're going to make up for all the heartache you've caused me, right?"

"I'll be glad to," he said.

Then he kissed me again. Hard. In his arms, I simply disappeared. Gone was the street. The houses. The earth. It was just me and Sterling. This could possibly be the best Christmas ever, despite all that had happened.

When we broke apart, I saw a man down the street out of the corner of my eye. Sterling and I both turned to him.

"Emma?" The voice sounded awfully familiar.

He came closer and a face emerged.

It was Nick Doyle. My ex.

CHAPTER EIGHTEEN

"Nick? What are you doing here?"

Nick's face was contorted in shock. And hurt. His dirty blond hair was combed neatly to one side. He wore a black designer winter coat, and he looked like he had just stepped out of a magazine spread.

"I came to see you," he said. "Aren't you going to introduce me?"

I noticed Sterling's jaw tense.

"This is Sterling," I said slowly. "Sterling, this is Nick."

Sterling put out a hand, but Nick didn't take it.

"You've moved on already? With him?"

I crossed my arms. "What's it to you?"

"I'm here because I wanted to see you and talk to you."

"About what? A simple phone call couldn't suffice?"

"I tried your cell phone, but you didn't pick up. Your manager told me where you were. Then I saw some tabloid pictures in the papers and got worried. Someone was trying to murder you?"

"The killer's been caught," I said. "I'm fine now."

He gave Sterling another steely once-over. "I can see that."

"We're over," I told him. "Yes, I am with Sterling. I have the right to move on. Just like you moved on with that Tara, right?"

Nick groaned. "You should know as well as anybody that tabloid stories are false. I went to Tom's birthday party, and some models were there. They're usually everywhere. We just took a picture together."

"It doesn't matter," I said. "I'm not going to interrogate you about what you did or didn't do. That wasn't the reason we broke up."

Nick sighed.

"I know," he said softly. "I messed up. I wasn't willing to fight for you. I took you for granted. Which was why I came here."

He reached into his pocket and took out a small box. He opened it. The most dazzling diamond ring shone back at me.

"I planned to propose."

My jaw dropped.

"But I see that you're busy with loverboy here."

With that, Nick turned away. I only stood there, aware that Sterling was watching me watch him.

RECIPE 1: HOT CHOCOLATE LATTE

Ingredients:
- 4 cups brewed hot coffee
- 1 cup half-and-half
- 1/4 cup chocolate syrup
- 1/2 teaspoon vanilla extract
- 2 tablespoons sugar

Stir everything in a pan over low-medium heat for five minutes or thoroughly heated. Serve with whipped cream!

RECIPE 2: ORGANIC HOT CHOCOLATE

Ingredients:
- Organic milk or unsweetened nut milk
- 4 to 8 pieces of high quality chocolate (preferably organic)
- Cream (optional)

Pour milk in a mug, leaving two inches for the chocolate. Heat this the milk on a stove. Meanwhile, chop up the chocolate. When the milk is warm (not boiling), throw the chocolate in. Pour back into the mug and enjoy! You can try this recipe with different flavors of chocolate, such as mint or orange, if you are feeling more adventurous.

RECIPE 3: HOT CIDER NOG

Ingredients:
- 1 cup apple cider
- 1 cup milk
- 2 cups half-and-half
- 2 large eggs
- 1/2 cup sugar
- 1/4 tsp ground cinnamon
- 1/8 tsp ground nutmeg
- 1/8 tsp salt
- 1/2 cup whipping cream, whipped

Whisk all ingredients except whipping cream in a saucepan over medium-low heat. Whisk occasionally until mixture thickens and coats a spoon (around 15 minutes). Top with whipping cream. Garnish with a cinnamon stick if you'd like.